Transforming Barcelona

Barcelona is perhaps the European city that has become most renowned for successfully "remaking" itself in the last 20 years. This image has been spread cumulatively around the world by books, television and internet sites, as well as by films and the use of the city in advertisements. This book is about the reality behind that image. It describes how the governors, professional planners and architects, and, in a wider sense, the people of the city and of Catalonia brought about the succession of changes which began around 1980.

Transforming Barcelona is made up of chapters by those involved in the changes and by local experts who have been observing the changes year by year. Particular emphasis is placed on the political leadership and on the role of professionals, especially architects, in catalysing and implementing urban change. Other chapters look closely at the way that the treatment of public space became a key ingredient of what has been termed the "Barcelona model". Critiques of this model consider from two different perspectives — environmental and cultural — how far the generally acclaimed successes stand up to full analysis. The book is fully illustrated and contains an up-to-date bibliography of writings on the city in English.

This book will be of particular interest to those studying or practising in the fields of urban regeneration, urban design and urban geography, as well as all those fascinated by the multiple dimensions of the current transformation of older cities. No other book is available in English that provides such extensive understanding of Barcelona from authoritative local participants and observers.

Tim Marshall has worked in and studied planning since the mid-1970s, and he has been visiting Barcelona since 1980. He worked in Barcelona at the start of the 1990s, researching the environmental change of the metropolitan region. Tim Marshall currently teaches at Oxford Brookes University in regeneration and European planning, and he regularly organises visits by English planning schools to Barcelona. The combination of these interests and experiences has given him a continuing insight into what has been going on in the city and in Catalonia as a whole. In particular, this insight reinforced the view that those best situated to explain the Barcelona experience are those on the spot, hence the approach taken in this book.

Other titles available from Spon Press and Routledge

The Enterprising City Centre
Manchester's Development Challenge
Gwyn Williams

Remaking Birmingham
The Visual Culture of Urban Regeneration
Edited by Liam Kennedy

Towards an Urban Renaissance
The Urban Task Force

Urban Future 21
A Global Agenda for Twenty-First Century Cities
Peter Hall and Ulrich Pfeiffer

The Chosen City
Nicholas Schoon

Cities for the New Millennium
Edited by Marcial Echenique and Andrew Saint

For further information and to order from our online catalogue visit our website at www.sponpress.com

Transforming Barcelona

Edited by Tim Marshall

Routledge
Taylor & Francis Group
LONDON AND NEW YORK

First published 2004 by Routledge
11 New Fetter Lane, London EC4P 4EE

Simultaneously published in the USA and Canada
by Routledge
29 West 35th Street, New York, NY 10001

Routledge is an imprint of the Taylor & Francis Group

© 2004 Selection and editorial material, Tim Marshall; individual chapters, the contributors

Typeset in Univers by Alex Lazarou, Surbiton Surrey
Printed and bound in Great Britain by TJ International Ltd, Padstow, Cornwall

All rights reserved. No part of this book may be reprinted or reproduced or utilised in any form or by any electronic, mechanical, or other means, now known or hereafter invented, including photocopying and recording, or in any information storage or retrieval system, without permission in writing from the publishers.

British Library Cataloguing in Publication Data
A catalogue record for this book is available from the British Library

Library of Congress Cataloging in Publication Data
Transforming Barcelona: the renewal of a European metropolis/edited by Tim Marshall.
      p. cm.
  Includes bibliographical references and index.
  ISBN 0-415-28840-1 (hardback: alk. paper) – ISBN 0-415-28841-X (pbk.: alk. paper)
    1. City planning – Spain – Barcelona. 2. Urban policy – Spain – Barcelona. 3. Barcelona (Spain) – History – 20th century. I. Marshall, Tim, 1950–.
HT169.S72B378 2004
307.1'216'094672–dc22

                                             2003027721

ISBN   0-415-28840-1 (hbk)
          0-415-28841-X (pbk)

Contents

Illustration credits	vii
Publication sources	viii
Contributors	ix
Acknowledgements	xii

1 Introduction   1
Tim Marshall

Part 1 What happened   25

2 Urban dynamics, public policies and governance in the metropolitan region of Barcelona   27
Oriol Nello

3 Behind Barcelona's success story — citizen movements and planners' power   47
Nico Calavita and Amador Ferrer

4 Governing Barcelona   65
Pasqual Maragall

5 Ten points for an urban methodology   91
Oriol Bohigas

6 The city, democracy and governability: the case of Barcelona   97
Jordi Borja

7 The planning project: bringing value to the periphery, recovering the centre   111
Juli Esteban

8 Public spaces in Barcelona 1980–2000   151
Nuria Benach

9 Public space development in Barcelona — some examples   161
Jordi Borja, Zaida Muxí, Carme Ribas and Joan Subirats, Jaume Barnada, and Joan Busquets

Part 2 Present and futures **173**

10 Barcelona's new projects 175
Barcelona Regional S.A.

11 The transformation of Poblenou: the new 22@ District 191
Oriol Clos

Part 3 Critical perspectives **203**

12 Urbanism, culture and the post-industrial city: challenging the
"Barcelona Model" 205
Mari Paz Balibrea

13 Changing course? Principles and tools for local sustainability 225
Enric Tello

Bibliography 251

Index 257

Illustration credits

The authors and the publisher would like to thank the following individuals and institutions for giving permission to reproduce illustrations. We have made every effort to contact copyright holders, but if any errors have been made we would be happy to correct them at a later printing.

Amador Ferrer   3.1–3.3
Barcelona 1979/2004. Del desarollo a la ciudad de calidad, Ajuntament de Barcelona, 1999   1.6–1.8, 4.2, 5.1, 5.2, 6.3, 9.3
Barcelona City Council   7.1–7.9, 9.6–9.8, 11.1–11.3
Barcelona Metròpolis Mediterrània, 15, 1990   1.9, 2.1
Barcelona Metròpolis Mediterrània, 44, 1998/1999   9.11, 13.3, 13.4
Barcelona Metròpolis Mediterrània, 59, 2002   13.1, 13.5
Barcelona Regional   10.1–10.3
Dinàmiques Metropolitanes a L'Area i la Regió de Barcelona, Mancomunitat de Municipis, Barcelona, 1996   1.3, 1.4, 2.2
E. Tello   13.2
Joan Trullen in *Criteris per a una Política Territorial de la Regió Metropolitana*, Consell Assesor d'Estratègies urbanes, Ajuntament de Barcelona, 2001   2.3
M. Balibrea   9.1, 12.1–12.3, 12.5
M. de Torres i Capell, *La Formació de la Urbanística Metropolitana de Barcelona*, Mancomunitat de Municipis, Barcelona, 1999   1.1, 1.2
N. Benach   8.1–8.4
T. Marshall   4.1, 4.3, 6.1, 6.2, 9.2, 9.4, 9.5, 9.9, 9.10

Colour section
Amador Ferrer   3, 4
Barcelona City Council, via Juli Esteban   2
Barcelona City Council   7, 8, 15
Barcelona Regional   10–14
Estudi de L'Eixample, Ajuntament de Barcelona, 1983   1
M. Balibrea   6
T. Marshall   5, 9

Publication sources

The following chapters have already been published elsewhere. Thanks are given for permission to republish here.

Chapter 2: reprinted with permission of the Woodrow Wilson International Center for Scholars. Originally published in Tuchin, J. S., Varat, D. H. and Blair A. R. (eds) (2002) *Democratic governance and urban sustainability*. Washington DC: Woodrow Wilson Center for Scholars. (English corrected.)

Chapter 3: Nico Calavita and Amador Ferrer, *Journal of Urban History*, 26, 6, 793–807, copyright Nico Calavita and Amador Ferrer 2001. Reprinted by permission of Sage Publications Inc.

Chapter 4: published under the title "Carta de Pasqual Maragall a Pujol i a González" in *Barcelona Metròpolis Mediterrània*, 37, 33–63, September–October, 1997. The section on finances is published here in the same reduced form as in the English translation provided in that edition.

Chapter 5: published under the same title in *Architectural Review*, September 1999, 88–91.

Chapter 6: published under the same title in *International Social Science Journal*, 1996, 147, 85–93.

Chapter 7: published under the title *El Projecte urbanistic. Valorar la perifèria i recuperar el centre*, by Aula Barcelona, 1999, Number 4 in the series Model Barcelona, Quaderns de gestió (see www.aulabcn.com).

Chapter 8: published under the title "Tres aproximacions a l'espai públic barceloní", in *Espais públics. Mirades multidisciplinàries*, edited by Rosa Tello I Robira, Portic, Barcelona, 2002.

Chapter 9: published within pages 137–77 in *L'Espai Públic: ciutat i ciutadania*, edited by Jordi Borja and Zaida Muxí, Diputació de Barcelona, 2001.

Chapter 10: published in part as *Barcelona New Projects*, Barcelona Regional, 1999 (parts updated by Barcelona Regional for this publication).

Chapter 12: published under the same title in *Journal of Spanish Cultural Studies*, 2, 2, 187–210, Taylor and Francis, 2001. (www.tandf.co.uk/journals/carfax/14636204.html)

Contributors

Mari Paz Balibrea lectures at Birkbeck College, University of London, where she teaches Spanish literature and cultural studies. She has written extensively on the work of the writer Manuel Vàzquez Montalban and is currently preparing a book on the role of culture in post-industrial Barcelona.

Barcelona Regional S.A was founded in 1993 by entities (Barcelona City Council, metropolitan bodies, etc.) and companies from the public sector to provide a common technical instrument in the field of urban planning, environment and infrastructure. Its main tasks include undertaking surveys, carrying out technical and economic feasibility studies, researching on urban planning, infrastructure and environmental issues, as well as the promotion, management and implementation of all kinds of infrastructure, urban planning, architectural and environmental projects.

Jaume Barnada is an architect, with a doctorate in architecture (Universitat Politecnica de Catalunya). He teaches in the School of Architecture of Barcelona and is Director of Land and Housing Policies in Barcelona City Council. He is the author of articles in various magazines and books, particularly on his areas of expertise in urban spaces and the strategies of urban renovation.

Nuria Benach teaches at the Department of Human Geography, University of Barcelona, and researches in the analysis of urban spaces in relation to consumption, identity and images.

Oriol Bohigas is an architect and planner, a founder and for many years director of MBM Architectes. He now works as an associate of the firm (now Martorell, Mackay, Capdevila and Gual). He was director of the School of Architecture of Barcelona in the 1970s, director of planning in Barcelona City Council 1980–1984 and councillor in charge of culture in Barcelona City Council 1987–1991. He is the author of many books, from *Barcelona entre el Pla Cerda i el Barraquisme* (1963) to *Cartes de la Baralla* (2003).

Jordi Borja is an urban geographer and sociologist, who divides his time between teaching, writing and consultancy. He was a senior councillor in Barcelona City Council from 1979 on, particularly involved in leading work on governance, whether in the decentralisation of the council in ten districts or in the preparation of the city charter. He has carried out teaching and consultancy in many countries, especially in Latin America. He writes a regular column for *La Vanguardia* newspaper. Books in English include *Local and Global*, with Manuel Castells.

Joan Busquets is an architect. He founded the Barcelona Laboratory of Urbanism in 1969 and was a professor in the Barcelona School of Architecture until 2002. He is now professor at GSD Harvard University, as well as guest professor at several European universities. He was head of the planning department in Barcelona City Council from 1983 to 1989. He has written several books on planning and design, and has worked on major projects in a range of cities, including The Hague, Rotterdam and Buenos Aires.

Nico Calavita is a professor in the Graduate Program in City Planning at San Diego State University. His research interests include growth conflicts, equity planning, affordable housing and comparative planning.

Oriol Clos is an architect, and is Director of Urbanism at 22@bcn, S.A., Barcelona City Council. He worked for many years in architectural practice on a wide range of building and planning projects, and was also, from 1990 to 2002, a teacher of urban planning in the School of Architecture of Barcelona.

Juli Esteban is an architect, specialising in planning. He participated in the making of the Barcelona General Metropolitan Plan, was director of the planning services of the Metropolitan Corporation of Barcelona, and is now director of the Cabinet of Urban Studies, Department of Planning, Barcelona City Council. He is also the author of planning text books, as well as of many articles, and he teaches at the School of Architecture of Barcelona.

Amador Ferrer, doctor in architecture and planning (Universitat Politecnica de Catalunya), works on urban projects and city planning, is professor of urbanism in Barcelona and other universities, has served in recent years as director of building (1988–1992) and associate director of urban studies (1992–1999) at Barcelona City Council, and director of urbanism at Badalona City Council (2000–2003). Since 2003 he has been coordinator of planning services at the Mancomunitat de Municipis for the Barcelona metropolitan area.

Pasqual Maragall was mayor of Barcelona 1982–1997, president of the Council of European Municipalities and Regions, and vice president of the EU Committee of the Regions 1991–1997, and then leader of the opposition (Partit dels Socialistes de Catalunya) in the Catalan parliament 1999–2003. Since December 2003 he has been president of the Generalitat, the Catalan regional government, as head of a three party coalition. He has a Masters degree in International and Land Economy (New School University in New York), and a doctorate in economics (University of Barcelona).

Tim Marshall has taught planning at Oxford Brookes University since 1991. Before that he worked as a professional planner in Birmingham and London. He has a doctorate on local planning for urban renewal in English cities in the 1970s and 1980s. He participated in planning and environmental research projects in Barcelona in 1990–1991. His current major research interest is in national and regional planning in Britain and Europe.

Zaida Muxí, a doctor in architecture (University of Sevilla), teaches at the Architecture School of Barcelona, and is the author of articles on design, architecture and planning. Her book *La Arquitectura de la Ciutat Global* will be published by Gustavo Gili in 2004.

Oriol Nello is a geographer. He was a researcher at Johns Hopkins University, and director of the Institute of Metropolitan Studies, Autonoma University of Barcelona from 1988 to 1999. He is the author of many books and articles, including recently *Ciutat de Ciutats* (2001) and *Aqui No! Els Conflictes Territorials a Catalunya* (2003). Since 1999 he has been an MP in the Catalan Parliament, leading on urban and territorial affairs, and in January 2004 he became a minister with responsibilities including planning in the new government of the Generalitat headed by Pasqual Maragall.

Carme Ribas is an architect and teaches in the Architecture School of Barcelona. In recent years she has worked on various architectural and planning projects in Barcelona, such as the Mercat Sant Antoni, the Baro de Viver housing estate and the remodelling of the Trinitat Nova neighbourhood, with the participation of the residents.

Joan Subirats is professor of political science in the Autonoma University of Barcelona, has a doctorate in economics, and is the author of several books in the field of public policy and politics, as well as being a regular contributor to the press and other media.

Enric Tello is professor of economic history, University of Barcelona, where he teaches a course on Economy, Ecology and Society. He has been involved in the preparation of the Local Agenda 21 in Barcelona, and he is involved in several citizens' campaigns for a less unsustainable city.

Acknowledgements

This book is the result of varying degrees of attention to the Barcelona phenomenon over more that 20 years. Very large numbers of friends and acquaintances have contributed to my understanding over this time. Barcelona is a good city to be interested in because the people are extraordinarily helpful and friendly, and I admit I have taken full advantage of this.

Because of this long engagement, I think the usual let-out is justifiable: I am going to thank these many people collectively and anonymously, now. Having said that, I must mention a very few people. I have learnt a lot from Amador Ferrer over the years, also from Josep Maria Carrera in Barcelona City Council. Recently, Joan Roca has provided very stimulating company, and has helped British students to understand the almost incomprehensible on the week-long visits, as Amador Ferrer did in earlier years.

Of course, I am very grateful to all the contributors to this volume, who agreed to their existing work being used here, or revised or wrote anew chapters. I hope the translation of their ideas to a wider audience proves satisfactory to all of them.

On the production side, in Oxford Brookes University I owe a great deal to Maureen Millard for organizing the text and doing wonders with retyping and scanning. Rob Woodward has been as essential in helping with the images. Many thanks to both. Other colleagues have helped less directly, and the School/Department has been an extremely supportive environment over these years. Field visits to the city since the start of the 1990s with staff colleagues Pete Edwards and Roger Mason have been consistently enjoyable. Increasingly, visiting Masters students have asked difficult questions, which always helps.

Above all, thanks to my close friends in Barcelona, without whom this certainly would not have happened.

Chapter 1

Introduction

Tim Marshall

As a city, is Barcelona particularly different from many others in Europe, especially in southern Europe? It is on the sea, and has been a port, as so many Mediterranean cities. Might we not expect a history of urban change and planning not unlike many cities in similar situations? There are after all many factors common, in very broad terms, to such a category as the "Mediterranean city" — climate, some geographical factors, certain similarities in the switchback of governmental change and war, a gradual process of modernization, particularly during the twentieth century, surges of urbanization, especially after the 1950s.

Yet Barcelona has in the past 10 to 15 years become the outstanding example of a certain way of improving cities, within both this Mediterranean world and in Europe, Latin America and even globally. The collection of articles in this book seeks to go inside this phenomenon, using the writings of those in Barcelona who have been involved in the changes or followed them very closely. The book has three objectives. First, and as the main task, it aims to give an account of what has happened in the city in relation to planning and urban change since about 1980. Familiarity with these elements should help readers to understand why Barcelona has picked itself out from the wider set of cities that might have been expected to engage in similarly creative ways — but did not. Some approach to explanation will therefore be made: not from a detached, outsider's perspective, but from the understanding of leading actors in the process.

Second, brief sketches of current programmes for the city are given. What new approaches and challenges are emerging? May we expect Barcelona to continue to innovate in interesting ways, to deal with the emerging new issues?

Third, there is an assessment of what has happened overall in these two decades. This does not claim to be a comprehensive evaluation. Readers will make their own judgements from the main text, as well as from their encounters with the city. These final chapters are designed rather to present assessments from cultural and environmental perspectives.

The task of this introduction is to give a short overview *from the outside*, to help readers to situate the accounts which follow. This overview will have five parts. First, I will outline the geographical and historical realities which lie behind and underneath the recent events. Second, I will comment on the traditions of urban planning in the city and, more widely, in Spain. Third, a set of "schematic facts" on the changes will be presented. The emphases I give to these "facts" are, of course, quite personal. Fourth, I will develop these summaries under some key dimensions of change, which may serve as headings for collecting together assessments from later chapters and from further analysis in the future. Finally, I will comment briefly on achievements and prospects, to give a personal take on the city's contemporary history.

Many of these issues can be followed up in the literature on the city, which is gradually growing. I have brought together what I hope is a reasonable selection of that literature in the Bibliography at the end of the book. This is made up *only* of publications in English, and so makes no claim at all to international general coverage, and certainly does not touch the more extensive writing in Catalan and Spanish.

Barcelona's history and geography

The most important features of the city's geography stand out on Figure 1.1. Climate helps. Unlike many European cities, Barcelona has neither very intense heat (just unpleasant humid heat in July and August, normally), nor cold — rarely sub-zero temperatures, just slightly cooler and wetter winters. Few other medium or larger cities, whether in Spain or elsewhere in Europe, have these qualities — except, perhaps, Valencia.

1.1
Physical location of Barcelona, shown on plan of 1934

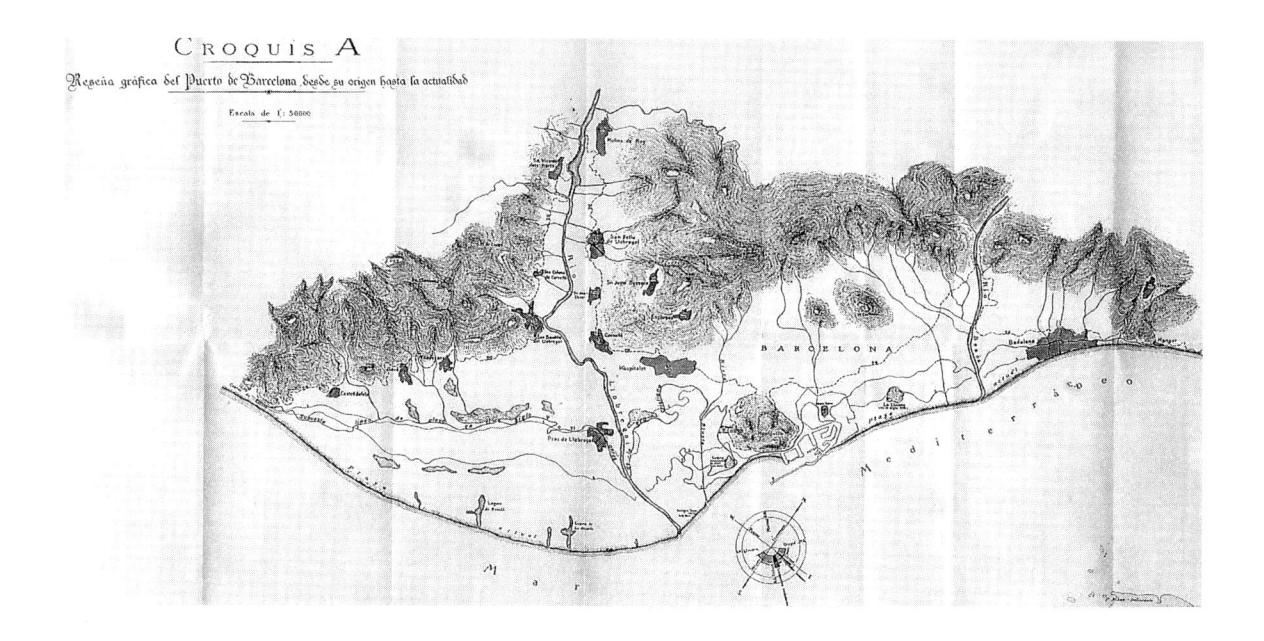

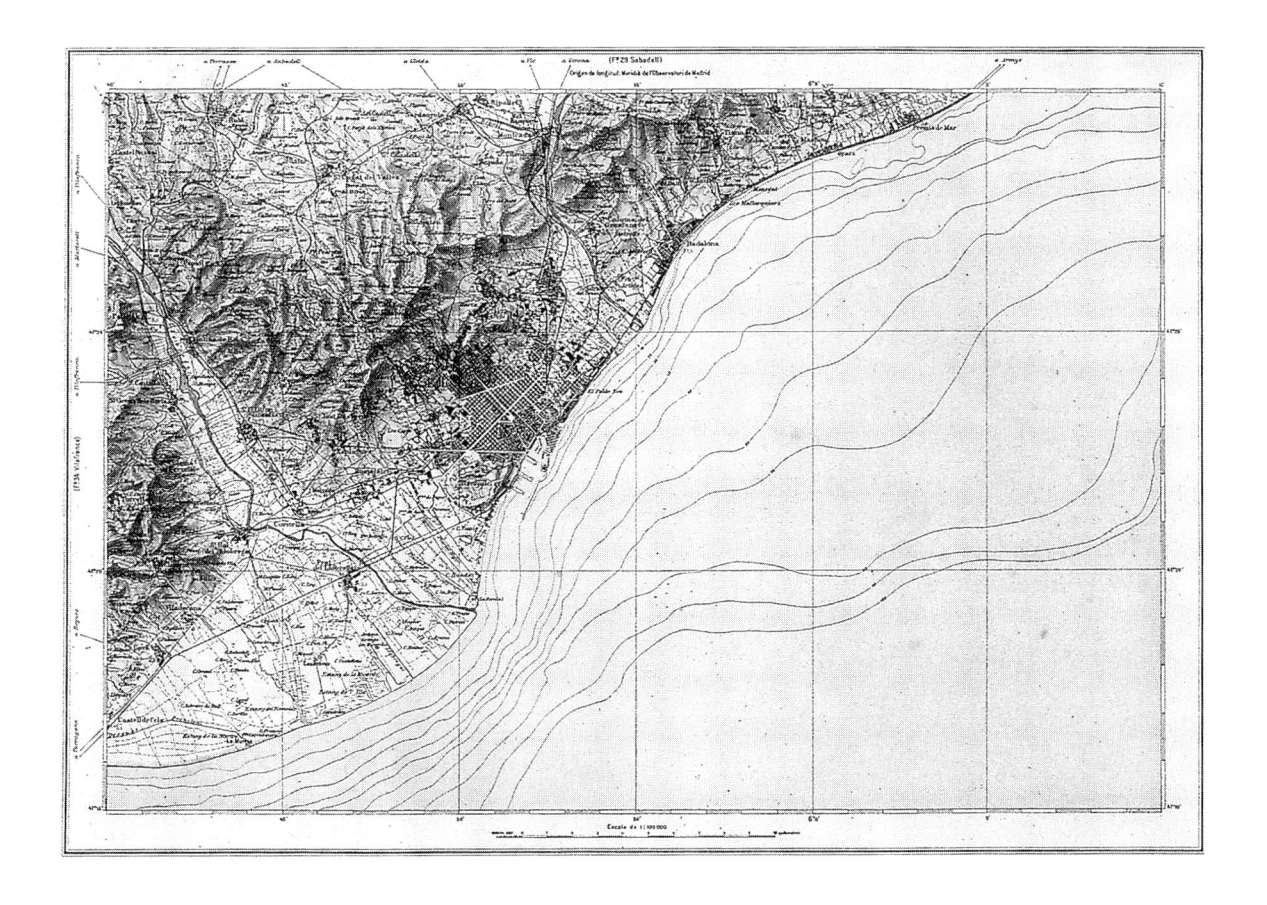

1.2
Coastal zone of Barcelona, 1923

The coastal location gives certain advantages — moderating the climate, easing communication for many goods, giving many leisure opportunities, serving as a sink for myriad wastes throughout the city's history. The position at a much wider scale has been emphasized as strongly conditioning in history (Vilar 1964). Barcelona is situated on a coastal corridor between France, Italy and the north, and the south and Africa. There is a thin strip along the coast itself from the city northwards, a range of hills behind that, and a wider inland depression, along which the Romans placed the Via Augusta. This location, in a kind of "march" or border country between zones of activity and civilization north and south, is seen as a vital factor in the forming of Catalonia, and with it Barcelona, as always the unchallenged capital (see Figure 1.2). Catalonia is seen as always taking much from the Europe to the north-east, and feeding these influences southwards. Its location has been one factor maintaining its cultural and political separateness, evidenced frequently over the period since the 1000s or 1100s when Catalonia may be said to emerge into a national history.

The nearest competing urban centres are far away, reflecting Spain's challenging geographical nature — Valencia 373 km south, Zaragoza 296 km inland, Madrid almost out of sight at 621 km. Southern France and Italy have often been as near as or nearer than most of Spain, in

INTRODUCTION 3

practical terms (both Marseilles and Bordeaux are slightly nearer than Madrid).

At a more local scale, the specific features of Barcelona's situation are that it is on a rather wide sloping site between the mouths of two important rivers, the Llobregat to the south and Besòs to the north. These rivers formed deltas, an important fact in the later Middle Ages when the deltas were brought under control and exploited for successive purposes (hunting, agriculture, water supply, waste dumps, industrial and airport sites, etc. — Marshall 1994). The rivers served as trade routes, channelling commerce into Barcelona. In the twentieth century, the rivers and deltas have been increasingly central in setting the opportunities and constraints open to planners or to urbanizing developers.

The other side of those constraints has been the balance of hills and flatter areas available, with the relatively inaccessible parts of Montjuic and Collserola setting limits to development, and then progressively becoming available as important leisure and ecological zones.

The importance of geographical and environmental history has been matched by that of economic and political history. Economically Catalonia, and Barcelona at its core, has been one of the two powerhouses of Spain in the past two centuries (with the Basque country). The accumulation of capital both within the territory and remotely via the Spanish empire (especially in Cuba up to 1898) allowed the continuing relative prosperity of the city and a continuing pressure to extend the urban area from the eighteenth century onwards. The bourgeoisie acquired sufficient wealth to drive the urbanization from that time on, and to import and develop a thriving material and artistic culture. Reserves of labour, first from the Catalan countryside and then from further afield in mid- and south Spain meant that lack of labour power did not constrain industrialization. The availability of cheap hydroelectric power from 1908 on, from the Pyrenees schemes, also favoured twentieth-century industry. A further essential boost was the formation of a chemical and oil complex from the 1950s in Tarragona, near enough but not so near as to blight the city's image. Nuclear power stations in the same zone of southern Catalonia followed in the 1970s. North African oil and gas were also easily importable. Barcelona's ecological foundations are varied.

This process of economic change through the nineteenth and twentieth centuries was not, of course, without its tensions, socially and politically. Class and interest struggle, at the workplace and in the city, was a more or less continuous feature of economic and political life, forming a central contribution to wider developments in Spanish politics. The epic all-Spain conflict of the Civil War 1936–39 overshadowed earlier and more localized struggles on the city streets, and prefigured the largely more pacific era of anti-dictatorship movements of the 1960s and early 1970s. The civil war defeat of Catalonia, and with it Barcelona, is a vital ingredient in the urban history of the period since then. This is one of the distinguishing features that must be at the forefront of any explanation of planning's vitality since the 1970s. The political nature of that force may be distasteful to some who expect design and regeneration to emerge seamlessly from vibrant and inspired creative professionals. Barcelona's experience has not been the sanitized story it is sometimes presented as. Donald McNeill's book (1999) on the city has given English-language readers much of the basis for a more sensitized understanding.

Another vital feature was the success of industrial development in the region, giving a base of confidence for tackling the challenges of urban stress; challenges that came with the period of rapid urbanization from the 1950s to the 1970s. Car ownership, motorways, second home ownership, higher living quality expectations, high rates of immigration, a larger local tax base, great expansion of higher education: all these laid foundations for the response made after 1980. Francoism's "great leap forward" after 1959, opening the economy to foreign investment, was essential to the catch up to near western European living standards. If the new city council in 1979 had started from a 1959 economic base, the options would have been utterly different, for good or ill (Salmon 1995, 2001).

The spread of development can be seen on Figure 1.3. The core municipality of Barcelona had successively annexed surrounding municipalities between the 1870s and the 1920s, so that it had space to control its own growth up to the most recent decades. It is true that in the second half of the twentieth century there was also massive growth in the "first ring" of municipalities, which virtually joined with Barcelona by the 1970s, forming a "real city" of over 3 million inhabitants. But the challenge of dealing with this wider scale was one to be addressed largely later.

This political issue of jurisdictions was very important, as students of large cities around the world are aware (Sharpe 1995). Barcelona became, through annexation, large enough to make its weight count, matching resources to the confidence which came with knowing that, quite clearly, it

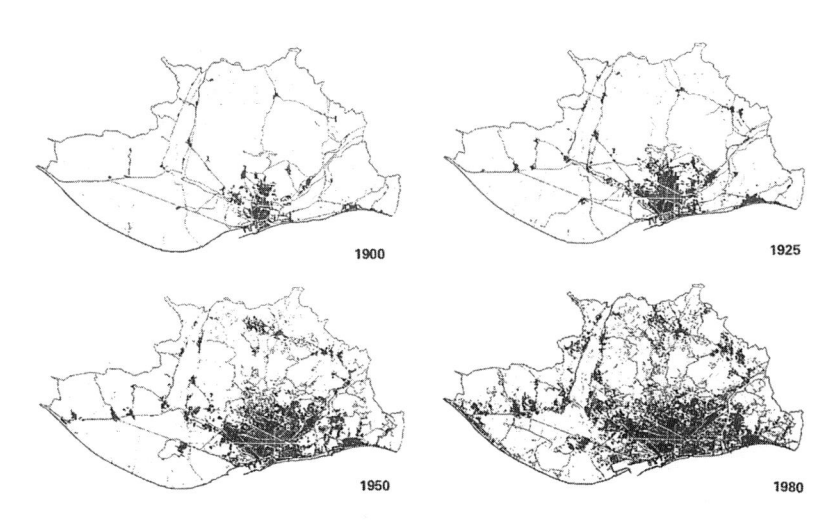

1.3
Urbanisation of Barcelona 1900, 1925, 1950, 1980

was Catalonia's capital and the challenger to Madrid in directing Spain's trajectory. This issue of political power runs through the planning history of the city for centuries, and gives a backdrop to every single big planning decision. The dynamics work at two main levels. The most important has always been between local forces in the city itself, and the Spanish state marshalled from Madrid. The subjugation of the city in 1714 meant over a century of being controlled by two overlooking fortresses and a ban on building outside the city walls. The release of that ban in the 1850s and the demolition of one fortress (the Ciutadella) allowed the "Extension" of the city (Eixample in Catalan, Ensanche in Spanish). Further bouts of central control in the Primo de Rivera regime of the 1920s and the Franco era 1939–75 left many further marks on the city's growth. The final period of relative freedom has been a time of autonomy, delightful for its freshness. But even that period has been full of a continuous negotiating with funding and regulating ministries in Madrid, mediated by party allegiances and nationalist differences — for example in the central government's funding of the Olympics schemes and the 2004 projects.

The secondary scale dynamic has been that between the city and the wider region, particularly the nation of Catalonia. Before 1980 this interplay was marked by the absence of any significant body at that level, except for the Metropolitan Corporation formed in 1974, made up of Barcelona and 26 neighbouring municipalities. This absence left the main tension as that between city and central government, with no one to represent the wider area, whether a metropolitan zone or the nationally defined territory of Catalonia or the even wider "Catalan lands" (with Valencia and the Balearic Islands). Since 1980 there has indeed been a powerful representative of Catalonia, the Generalitat, one of the 17 autonomous regional governments given major powers under the semi-federal constitution of democratic Spain. This body, run from 1980 to 2003 by the centre-right Catalan nationalist party Convergència i Unió, led by Jordi Pujol, abolished the Metropolitan Corporation in 1987 and sought to exercise strong influence over planning throughout Catalonia. It succeeded in this to a greater extent outside the metropolitan area than inside it. In Barcelona, and nearby, its influence has been rather one of impeding major efforts to plan long term and coordinate, than giving positive leadership. However, the Generalitat has broadly cooperated with Barcelona City Council in concrete projects such as the Olympics and the 2004 Forum.

The nature of Barcelona's planning tradition

As readers of the book will detect (especially in the chapter by Esteban), there are several contenders as to who were the key agents and what were the essential techniques involved in Barcelona's transformation. But most are agreed that one ingredient has been an ability to plan. In Spain, there has

1.4
Plans for Barcelona:
(a) Cerdà 1860
(b) Jaussely 1905
(c) Rubió i Tudurí 1932
(d) Macià (Le Corbusier/GATCPAC) 1934
(e) Pla Comarcal 1953
(f) General Metropolitan Plan 1976

long been a system of forward planning of towns and cities, called "urbanismo" (urbanisme in Catalan). This is particularly well entrenched in Catalonia, especially because of the legacy of one of the great founders of planning, Ildefons Cerdà (Magrinyà and Tarragó 1996, Soria y Puig 1995, Ward 2002). Cerdà is a strong candidate for the title of the first modern urban planner. An engineer by training, he also studied the society he was addressing with major statistical surveys. Politically, he was a utopian socialist, though one ready to get his hands dirty in the business of planning and trying to implement the expansion of Barcelona. His plan of 1857 for

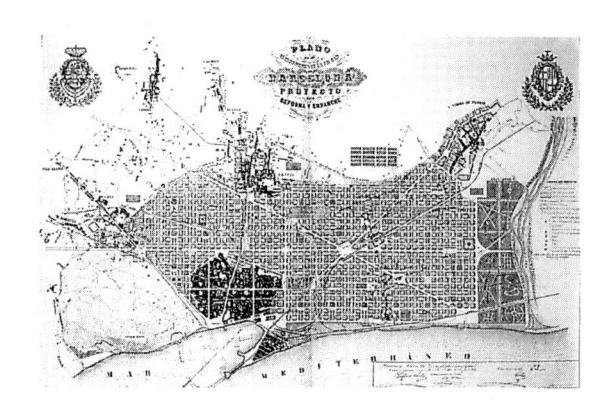

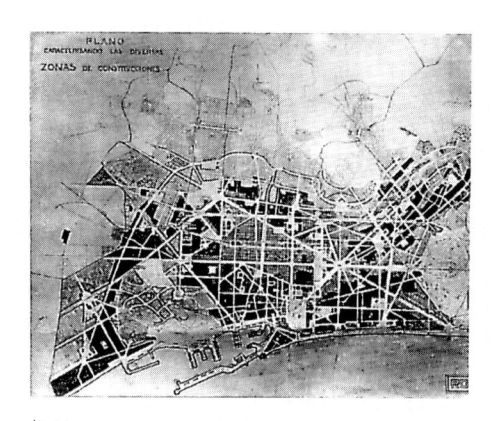

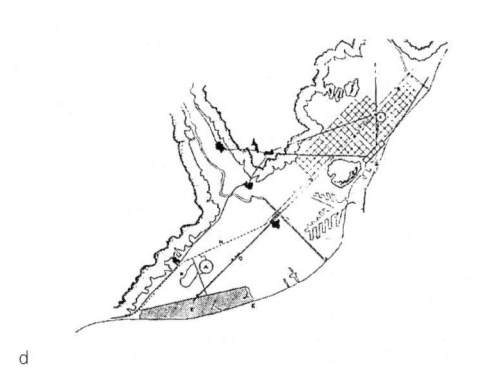

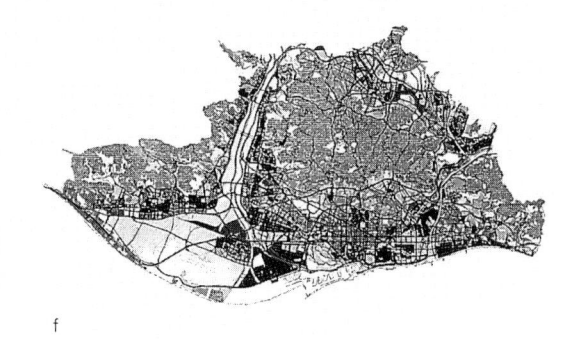

INTRODUCTION 7

the city remained the legal plan until 1953 (colour plate 1). The plan was often criticized in the succeeding century, but formed the foundation for a series of subsequent exercises, the most important being those of Leon Jaussely (1903), the modernists in GATCPAC (1931–33) and the Franco era plans of 1953 and 1968, which tackled the wider metropolitan region as much as the city itself (Figure 1.4 shows these plans).

This tradition of making plans, and to varying extents implementing them, gave legitimacy to planning activity in the city. Often the city was ahead of developments in Spain generally; the first coherent national planning law was passed in 1956. Planning was carried out by the physical design professions, engineers and architects — especially in recent years overwhelmingly by architects, many with supplementary training in planning (urbanisme) within their architecture degrees. These architects cooperate closely with other key professionals, above all economists, who have to check that the complex funding arrangements for infrastructure will succeed. The writings of many planners in Barcelona show a strong awareness of the tradition they inherit. It would be interesting to contrast this with other national traditions of planning. I suspect that in Britain there have been more frequent breaks in the history than in Catalonia, which probably weakens the activity and discourse of planning.

As Calavita and Ferrer show in Chapter 3, the architects became radicalized in the movement against the dictatorship and were essential drivers of the planning and implementation of urban change in Barcelona after 1975. Some of the politicians came from similar backgrounds, particularly Pasqual Maragall, Mayor 1982–97. He was a graduate in economics and law, who studied and lectured in New York and Baltimore in the 1970s, and completed a doctorate on land prices in Barcelona in 1979. Many of the councillors came from the anti-Franco residents and trade union movements, and so brought similar values to the decisions on planning after 1979.

The successes of the 1980s did not come without their frictions and professional rivalries. Moix (1994) shows how difficult it was at times to get the proud profession of engineers to work with the insurgent architects, brought into powerful positions by Oriol Bohigas, chief architect/planner 1980–84. Some sort of peace did not break out until towards the end of the 1980s, when the two professions began to appreciate each others' skills, especially in the overlapping zone of treating public spaces (invariably next to roads).

The involvement of politicians was often quite deep, even in the detail of schemes. Moix describes cases where Maragall was effectively the prime mover, as in the commissioning of Richard Meier to design the Museum of Contemporary Art (MACBA), or in his interventions over the design of the Hilton in the Diagonal.

Many other features of the planning tradition stand out, and will appear in several contributions in this book. They include the insistence on giving physical and visual form to schemes, normally beautifully presented —

increasingly, of course, since the late 1980s through computer graphics. The skills developed in dealing with open spaces and streets come out of the same inheritance, from Cerdà onwards. Plans for the broad structuring of the city (especially by transport arteries) are accompanied by design at often a very detailed scale. Some difficulty in dealing with or incorporating social and economic dimensions goes with the physical emphasis, though this is by no means always the case. Examination of a book of magnificently presented research like that by Mancomunitat de Municipis (1996) shows the strength of much economic and social analysis, and how this is informing planning. This has been one factor in the emergence of "strategic planning" in the late 1980s. This seeks to guide forward planning for the city as a whole more comprehensively, and ensure that locally oriented urbanism responds to the strategic plans emerging from these exercises (see Associació Pla Estratègic Barcelona 2000 published in 2000, Santacana 2000, Borja and Castells 1997, Marshall 2000).

In short, Barcelona's planning successes are incomprehensible without taking into account the technical and intellectual strength of several generations of sophisticated planners, but especially those born in the 1940s. This was a generation well versed in international planning experience, especially from Italy, Britain and France. What is equally noteworthy is the willingness of this generation to debate their views and activities in a wide range of forums in Barcelona and beyond, and to write about their experiences. This book pays some testimony to this effort (several chapters being by such architect-urbanists), but this is only the tip of an iceberg of much wider dissemination, in collaboration with geographers, economists and politicians. There are several forums for this discussion, including most importantly the professional body, the College of Architects, along with other bodies such as the Catalan Society for Territorial Planning (SCOT) and the university architecture schools. A very short list of books in Catalan or Spanish on Barcelona's planning would include Bohigas 1985, Busquets 1992, Carreras 1993, Borja and Castells 1997, Nello 2001: could other relatively small countries match this outpouring of publicly available debating on planning?

Some schematic facts

1   Barcelona municipality has seen massive physical change since 1980. This is normally depicted as progressing in waves, from smaller (squares, etc.) to medium (parks, etc.) to larger (1992, etc.). The pattern may be more complex. Overall the building work has surely not matched the 1960s or 1970s, but a large quantity of hectares has definitely been affected, and in a wide range of ways (building types, building uses, open space types, infrastructure changes, ranges of public facilities and so on). The improvements made to public spaces

have been admired around the world, as have some of the built schemes (such as the "Five Blocks" housing of the late 1990s in Poblenou).

2   This massive physical change has affected the city region quite unevenly. If the new built environment was mapped by contours we would see, by 2004, mountains along most of the coast from the Port Vell to Badalona, significant hills clumped around many points in Barcelona muncipality (C. Tarragona, Glories, RENFE Meridiana etc.) and a few in surrounding municipalities (Fira 2, the airport, etc.), and finally a scattering of more diffuse and lower raised points over the conurbation, in the Barcelona Eixample and elsewhere, as well as some linear ridges, above all the ring roads (and the mountain of Collserola would be flat!). The conurbation's centre is likely to be elongated towards the north-east, with the coastal projects (and eventually the high-speed train station at La Sagrera).

3   This change has been guided, as noted above, by sophisticated urban(ist) professionals, particularly architects and economists, employed largely by public agencies. Creativity and commitment has been notable. Expertise on areas of social and environmental concern has been supplied rather more by pressure from within society, and within political parties, than by paid professionals.

These professionals developed techniques of working, whether at the local level (for streets, all kinds of public spaces, all sorts of facilities) or at the strategic level (in the now much-exported city strategic planning model). The development and maturing of these skills is a feature of the city's change, however it is evaluated.

4   The wider metropolitan area has seen equally massive change, affecting more hectares in total, though of course in a large number of programmes. These physical changes have occurred with some elements of coordination, both within each municipality and from the (ex) Corporation, Mancomunitat and Generalitat, but the wider control has been relatively weak, compared with that exercised within the core municipality. But planning control of varying kinds has been the norm throughout the region.

5   Alongside and partially related to this physical development of the territory has been extensive change of the economy and the society, and of the environment. That is, the use of that physical infrastructure that is Barcelona city and region has changed significantly, often within the same physical envelope. Despite the massive range of analysis on these changes, it is not easy to establish the links between the physical and spatial changes and these parallel transformations. Some obvious changes are detailed below.

– Shifts in economic sectors, usually summarized as terciarization.
– Shifts in economic locations — creation of a new regional industrial ring, of large new office complexes, of new leisure and tourist complexes.

- Shifts in types of work, and of workers.
- Shifts in travel patterns, especially an explosion of commuting and leisure travel.
- A new period of immigration, in the 1990s, though on smaller scales than that from southern Spain in the third quarter of the century, and with different urban destinations from those earlier arrivals, and very little resort to self-build settlements.
- Higher levels of money incomes, for most — with relatively limited increases in social and spatial polarization (compared with some European countries) according to analyses in the late 1990s.

The 2001 census results allow us to pin down some spatial dimensions of these changes. Figure 1.5 and Table 1.1 show clearly how Barcelona has become a "shrinking city" in population terms, and that the same phenomenon was affecting the metropolitan area as a whole. Only immigration from abroad was doing something to reduce this shift, and ensure that the province of Barcelona as a whole saw a slight gain in population in the 1990s.

The net extent of these changes is not so easy to assess, overall. The emphasis of analysts has, understandably, been on change. But continuity may be just as definite.

6 A vital part of the reality has been the wider context within which the city region functions. The catch words here — Europeanization, globalization, a different relationship to the rest of Spain — help up to a point, but we understand them through the lenses of successful analysts and

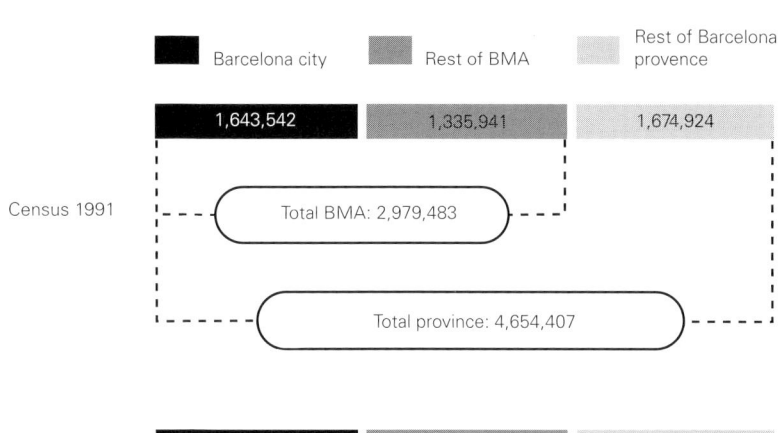

1.5
Increase of Barcelona's population 1991–2001 (author's elaboration of census data) (BMA = Barcelona Metropolitan Area, the 27 municipalities in the Mancomunitat de Municipis de Barcelona)

Census 1991

Barcelona city	Rest of BMA	Rest of Barcelona provence
1,643,542	1,335,941	1,674,924

Total BMA: 2,979,483

Total province: 4,654,407

Census 2001

| 1,503,884 | 1,329,465 | 1,972,578 |

Total BMA: 2,833,349

Total province: 4,805,927

Table 1.1 **Balance of residents 1991–2001 (includes those born abroad resident in 1991)**

	Total for province	Total for Barcelona Metropolitan Area	Barcelona	Rest of Barcelona Metropolitan Area	Rest of province
Natural and migration gains	489,602	270,800	138,767	132,033	361,299
Natural and migration losses	–596,059	–584,400	–387,276	–197,124	–154,156
Net balance of residents	–106,457	–313,600	–248,509	–65,091	207,143
Percentage of population 1991	–2.3%	–10.5%	–15.1%	–4.9%	12.4%
Born abroad (net increase 1991–2001)	257,977	167,466	108,851	58,615	90,511
Percentage of population 1991	5.5%	5.6%	6.6%	4.4%	5.4%
Total balance 1991–2001	151,520	–146,134	–139,658	–6,476	297,654
Percentage of population 1991	3.3%	–4.9%	–8.5%	–0.5%	17.8%

Source: INE (National Institute of Statistics)

journalists, and especially in the political ferment of Barcelona politics, through the pronouncements of political leaders. The catch phrases are reality, but may not match at all precisely material, lived change.

Barcelona's governors have no doubt often in the past related their projects to these wider changes in the world, as they saw them, and those of the last two decades have been no exception. Generally, they have "swallowed" the common story of "globalization" and seen their prime role as being to adapt to its perceived inevitable pressures. It is useful to identify this as a political choice, however we may evaluate it.

Four dimensions of change

Let us cut up this activity another way, viewing four different dimensions. This will help to explore the varying factors in play in the big changes.

Treating public spaces and buildings
This has been the area of intervention par excellence, at least as viewed by most visitors to the city, as well as for many citizens. It includes:

- new roads, and redesign of existing roads;
- new public spaces — squares, beaches, promenades (rambles);
- new public buildings, most delivering public services (libraries, museums, sports centres, offices, community centres, health centres, schools, markets), some doubling up as tourist venues;
- new infrastructure — many expanded or completely new facilities, whether the large storm drains, rail extensions or diversions, stations, airport, street lighting, telecommunications.

This effort represents large public investment, mostly with public funding and producing publicly managed and owned facilities. This may be, conceivably, the last phase of such investment. But it is another core ingredient of the Barcelona achievement. Such integrated achievement as is visible in many Barcelona neighbourhoods would be difficult to achieve with the now familiar jungle of public–private partnerships with which regeneration specialists have to struggle in most developed countries.

Often, observers have remarked on the quality of design across these programmes. Some exploration of this theme has uncovered the roots of this ability to concentrate on detailed implementation, with the development of a "design culture" from the 1950s, with one eye at least on Italy (Narotsky 2000). Discussion has also focused on the public space component of the programmes, drawing out a lesson on democratic interaction on the streets and a post-modern slant on the encounter of diversity that this enables (Borja 2000, Balibrea this volume). Streets are for all, whatever income, colour of skin or age a person may have.

Such an understanding chimes with the long-standing progressive tradition of wide parts of Barcelona society, including the strain of anarchism, in education and art, and the generation since the 1960s of powerful theatrical groupings: la Fura dels Baus, els Comediants, els Joglars being three alternative theatre troupes now feted around the world. The tradition, common to Mediterranean countries, of frequent use of streets and other spaces for festivals and parties, reinforces the willingness of politicians to invest in street quality and to sponsor regularly such festivals at the level of the city and each neighbourhood. Unthinking transfer to countries without such tradition needs very careful thought.

Economic dimensions of the urban changes
FINANCING THE PROGRAMMES
I have already commented in part on the public inputs to Barcelona's programmes. Private finance has been very important in many areas as well, and this has sometimes generated significant tensions and debates.

The most famous of these was that about the use of the Olympic Village. Maragall had promised that much of this housing (at least 50%) would be available for social housing. Subsequently, it became clear that funding was not available to achieve this. The whole project was handed over to private developers, and the area became one for up-market owners and investors. After 1992 the pressure has been on to produce the promised units for those on lower incomes, with success in other schemes in that area of the city (the Five Blocks scheme). But this has only been possible where public authorities could subsidize through their existing land holdings. Barcelona suffers the same problems as other cities in this respect. Only where housing finance (provided by the regional government) has been prioritized have social goals been achievable. This has mainly been in rehousing schemes, either in clearance areas of the old city, or in clearance of blocks of older public housing or private housing affected by aluminosis (concrete decay), in the outer suburbs.

Private finance has been successful in certain profitable sectors — the modernization of telecommunications infrastructure, up-market housing, the rapid increase in hotel provision, the booming of restaurants and bars, the creation of four or five major new shopping centres. Different opinions exist on these latter newcomers to Barcelona's life, challenging the dominance of more traditional neighbourhood provision, including smaller supermarkets and municipal markets. It is possible that neither city council nor regional government (with a final say on the creation of major retail provision) will want to approve more schemes after those recently completed. But those now existing may well hit local provision heavily.

It has been conventional since the early 1990s, in the UK at least, to couch discussion of this kind in terms of "partnership" — generally between governments and the private (profit-making) sector. A very clear description of Barcelona's work in these terms, but retaining an understanding of the trade-offs and risks involved, is provided by Francesc Raventós, one of the economists supporting the city's efforts in strategic planning and elsewhere during the 1980s and 1990s (Raventós 2000). Sections at the end of Chapter 10 provide some details on the funding of the current (2004) investment round, and of the Olympics.

Raventós discussed the varying balances and approaches used in six different spheres of public operation — the urban visual improvement scheme (Barcelona posa't guapa — make yourself beautiful), the Olympics, the recuperation of the Old City, the Plan 2000, international economic promotion, and the investment company formed to invest in industry. One of the factors he stressed was the importance of public sector leadership, and especially the need to commit municipal funds to catalyse action. As often in Barcelona, this may have a rather "old left" ring to it. But the most recent initiatives for the 2004 Forum and the 22@ area (described in Chapters 10 and 11) show that this approach is by no means of the past, although the emphasis has undoubtedly shifted as municipal finances and ideologies have changed.

In Britain and other north European countries, much of Barcelona's efforts since the late 1970s might have been presented in terms of the re-use of "brownfield" land. Old industrial sites have, after all, been the key source of development land since the early days, and partial or complete ownership by the municipality or other public bodies has been one necessary condition for many, perhaps most, of the achievements of the city. This relatively simple lesson about the importance of public ownership perhaps bears repeating, at a time when, again in the UK at least, the struggle to develop remaining brownfield sites and to avoid taking farmland is at the centre of planning debates. In fact, in Barcelona it may be precisely the lower amounts of publicly owned land in the next phase of intervention, in the 22@ area of Poblenou, which is behind the increased conflict with local residents. In this area the calculus of social gains/private profits is less conducive, it would appear, to public objectives. This has led the council to make plans that allow rather intensive development, including high rise schemes. Similar recent moves backed by the Mayor of London, who also, and in greater degree, lacks land reserves and public funds, suggest that this is not a situation unique to Barcelona.

IMPACTS OF THE PROGRAMMES

Most of the programmes for which Barcelona is renowned, certainly up to the mid-1990s, were not conceived with the aim of boosting the city's economy, or its global competitiveness. Many have, however, done precisely that. Making the central area, particularly the port and the old city, more attractive has allowed tourism to flourish in that area, with the beaches, provided with locals in mind, giving an extra attraction. A functioning public transport system is essentially for residents, but helps to make visitors' experience much more pleasant. It is no doubt this trick, of having made improvements for locals "pay" internationally, which has impressed many city managers around the world.

More widely, the good reputation that the city has obtained appears to encourage investment generally in manufacturing, research and services by multinationals. This is, at any rate, the interpretation that economists place on the city and region's success in this respect since around 1990 (Trullen 2001). It is presumably here that the major investments in motorways, the airport, industrial land provision around the region, and that proposed in the port, is paying off.

The 2004 Forum was designed much more consciously with wider economic objectives in mind. Those involved with the project in the city hoped that this further push would bring the city's reputation and functioning onto a new plateau, without at the same time leaving a problematic burden of debt. Evaluation after the event will probably be as difficult to do as after the Olympics. In the meantime, Chapter 10 by Barcelona Regional gives us some idea of the amounts of money and issues involved.

The politics of change

This book is primarily about change within the municipal boundary of Barcelona, even though such change is in many ways inseparable from wider processes, particularly in patterns of employment and residential decentralization. The municipal government is evidently the central actor in what has happened since it was given democratic legitimacy following the 1979 elections. We need to understand several aspects of this political leadership.

The dominant political party has been the Partit dels Socialistes de Catalunya (PSC), the Catalan affiliate of the Partido Socialista Obrero Espanol (PSOE). It has lost no election since 1979 (at the time of writing), an almost unprecedented feat in Spain, only matched amongst cities by La Coruña in Galicia. Socialist or Communist control, which was almost universal in the cities in 1979, has normally given way to centrist or right wing parties. Always, the Socialists have had to make pacts with one of the smaller parties, generally the Communists, or their successors Iniciativa per Catalunya, and in the 1990s Ezquerra Republicana, a historic centre-left party favouring Catalan independence. But the Socialists have normally had around 16 seats to the 3 or 4 of the smaller parties. The two right wing parties of Convergència i Uniò and Partido Popular have always been rather weak in the city, even though until 2003 one ran the regional government and the other, between 1996 and 2004, the central government.

The Socialists have had three leaders of the council. The first was Narcís Serra (Figure 1.6), a heavyweight locally who became a minister (finally deputy prime minister) in the PSOE Madrid governments 1982–96. The second was Pasqual Maragall (Figure 1.7), who dominated municipal politics until his resignation in 1997. He was from a well-known family of the Barcelona bourgeoisie, grandson of perhaps Catalonia's most famous modern poet, Joan Maragall, and already very familiar with how the council worked (or did not), as an employee through much of the 1970s. Spanish mayors are not elected directly, but once installed as leader of the Socialist group, he could wield considerable powers of decision making and patronage. Such power was naturally reinforced by successive victories in elections in 1983, 1987, 1991 and 1995. He had a strong interest and expertise in planning, which became one of the cornerstones of the council's drive. However, it was by no means the only major emphasis. Part of the planning success doubtless lay in the ability to combine urban development with policies on culture, economic development, transport (through control of the metropolitan transport body which ran metro and buses), to name the most important. Nevertheless there were difficulties

1.6
Narcís Serra, Mayor of Barcelona 1979–82

1.7
Pasqual Maragall, Mayor of Barcelona 1982–97

1.8
Joan Clos, Mayor of Barcelona 1997–

in achieving full integration, as the council remained dependent for many policies on the regional government — to varying degrees for education, housing, social services and transport. Maragall's ability to make deals with Pujol and with Gonzalez (in Madrid) was vital to the council's achievements. The same point would surely hold if we were examining the achievements of urban leaders in New York, London, Paris — or Madrid itself.

This need for coordination was particularly vital in the Olympics phase, when the meeting of deadlines was by no means guaranteed, as a Barcelona journalist's blow-by-blow account of these years revealed (Moix 1994). It was necessary to set up a top-level working group at the end of 1988 to drive forward the works and exercise financial control, with a special representative brought in by Madrid to look after its budgetary interests, and a Generalitat member, too. The group was chaired by Maragall and met fortnightly.

The third leader was Joan Clos (Figure 1.8), groomed by Maragall for a year or two before 1997. He had made a success of renewal in the old city, and was able to retain Socialist control in 1999 and in 2003. As a medical doctor, he did not have the background interest or expertise in planning, and there was a sense amongst professionals that planning matters were downgraded after Maragall's departure. But Maragall had, before leaving, pushed the commitment to a new "big event", the 2004 Forum, and this legacy meant that the council had a certain programme of work laid out for at least seven years to come.

There is no doubt that the quality of mayor has mattered in Barcelona. The support of other key politicians has also been important. Amongst these may be mentioned Jordi Borja, an academic and political activist, originally in the Communist party (PSUC), but by the 1980s an independent. He took over the job of decentralising the organization of the council, a vital part of the reform of the council's structures. From 1986 there were ten "mini town halls", each with a local district council made up of the councillors for that part of the city and a district leader. District offices gradually took over all front-line services, although specialist services have remained in a strong central core. This localization of services was seen as key to both serving citizens and allowing wider participation. It was a natural outcome of the force coming from the residents associations, which were still flourishing in the early 1980s. Borja added his own emphases as a theorist of popular control, a sociologist who had studied with Manuel Castells in the 1960s in Paris. He particularly supported the programme of building community centres in each neighbourhood. Subsequently he was to move on to thinking about city management and planning more widely, as evidenced in the publication of a book by himself and Castells (1997).

A very wide semi-technical, semi-political elite supported these politicians through the 1980s and 1990s, part of the same generation and with broadly the same values — progressive, egalitarian, gradualist, as interested in economic and social as cultural dimensions of the city's renewal, a mainly middle class grouping, products of the university expansion since the 1960s. As described, they provided in particular the cadre of expertise in the planning sections of the council. Unlike some smaller Spanish cities, Barcelona has had a fairly well staffed planning service, and could draw in extra expertise from the universities. It was nevertheless necessary, when Oriol Bohigas (Figure 1.9) arrived as planning director in 1980, to bring in a phalanx of very recently graduated architects to get the new programmes moving (Moix 1994). The same skilled personnel base applied in the strategic and economic planning which emerged in the late 1980s. The first "Economic and Social Strategic Plan Barcelona 2000" was prepared by drawing liberally on the skills of sympathetic academic expertise, as well as, to a lesser degree, on staff in the trade unions and employers organizations (Marshall 1996).

1.9
Oriol Bohigas, architect, director of planning of Barcelona City Council 1980–84

Reference to the employers bodies is a reminder that the council's project was far from disturbing to most of the "normal powers" in the city. Although business may have generally been more sympathetic to the right wing parties, they were, at least in part, successfully brought inside the council's project with the Olympics, Barcelona 2000 and other initiatives. It can be argued that the council leaders forged a relatively successful hegemonic project during this period. Academics may argue whether this is best described in the political science language of "urban regime" (Lauria 1997): an interest-based, enduring coalition, to an extent cross class, which delivers a city elite's aspirations as well as a broader programme. At any rate, up to the time of writing there has been no effective counter project. It would seem that there has been an ability to open to new currents, whether by adopting the language and practice of partnership with business in the 1990s, or by pressing green credentials, against ecologists' criticisms, in the same years: for example by the creation of a "sustainable city" division and programme in the late 1990s.

Is the council as open to public participation and influence as it was 10 or 15 years ago? Many residents groups in the 1990s would say not, as evidenced in the consistent criticisms in the residents' federation newspaper "La Veu del Carrer" over this period (and see Huertas and Andreu 1996, FAVB 1999). Probably the change of approach in the council to more dependence on private funding has brought a shift of attitude, narrowing the range of matters on which the council wishes to negotiate with local groups. However, there is probably still a broad commitment to discussion on many issues, where the council feels it has room to manoeuvre.

Planning therefore rested on political foundations, specifically the force of the social movements of the 1970s, the careful building in the 1980s of a cross-class hegemonic coalition by Maragall and his associates, and the absence of any other really challenging political forces in subsequent years. Lest this summary and much in this book makes the path look too smooth, it should be emphasized that there were many tensions, throughout the period. Luck, careful management and many other factors kept the programmes more or less on course, but this was by no means inevitable. Many criticisms can be made of what Barcelona has done or not done, and these have been as much caused by these political dynamics as the widely acknowledged successes.

Social dimensions of planning

As I have explained, urban development and planning has been multi-faceted over these years, moving forward on many fronts at once. On all of these it can be argued that there have been gains for a broad swathe of the city's inhabitants — in public spaces, in public facilities, in infrastructure, in housing, in providing jobs. It is difficult to make any broad assessment of such an effort over more than two decades. Whilst some efforts were made to evaluate particular elements of the programmes (the Olympics — Brunet 1995, the ring roads — Riera 1993), a wider coverage would be hard. From the left there have been major criticisms. One was at the time of the Olympics, when it was argued that the investment style and practice of the council was not so different from that of the Franco years, with authorities and developers working hand in glove in ways that might have been close to corruption (Moreno and Vázquez Montalban 1991). Other criticisms in more recent years have detected a drift towards business-friendly planning, marked by developments like the high rise US style development at Diagonal Mar (see Chapter 9; also McNeill 2003). The debates about the highly commercial development of the old port in 1989–93 had had the same concerns, seeing Baltimore's projects of leisure schemes and shopping malls unthinkingly transferred. Since 2000 there has been sharp controversy about the plans for the 22@ Quarter in Poblenou. This scheme, which intensified development for both business and housing use well above Metropolitan Plan norms, was opposed vigorously by residents in the area, and in 2002 the council backed down to some extent, reducing intensity and the height of proposed tower blocks.

More generally, there has been persistent criticism of the council's inability to tackle housing needs, for those who cannot afford the spiralling cost of paying mortgages or rents in the city since the late 1980s. As mentioned above, this is not a failure limited to Barcelona amongst western cities. However, it is the one on which there has been perhaps most debate in recent years, and where the council's record may be most clearly weak. If the Barcelona model includes such a central failure, then evidently that model needs careful examination before being given undiluted praise.

More generally, there has been a view that Barcelona is gradually becoming a city more suitable for rich people (including tourists) than for ordinary citizens. The cost of living in the city has certainly been a factor in the continuing exodus of people, which has brought its population down to one and a half million from about 1,800,000. This decline might continue, though probably at a lesser rate. This throws up as a major issue the planning of the wider metropolitan region, because many of these people still depend economically, or educationally or in other ways, on the central city. But it is precisely this wider scale of change where limited progress has been made in the last decade, as Nello discusses (and as I have discussed elsewhere, 2000). This shifting in the social composition of the city region is regarded by many commentators as the key issue for the present. Whilst it is probably exaggerated to see the planning processes underway as creating a city for the rich, rather than the thoroughly cross-class city of the past, there are surely major questions arising about future social change.

Achievements and prospects

Few argue against the scale of achievement of the first 10 or 15 years of urban work in the city. This remains a laboratory of successful practice that will continue to repay study, perhaps for some years. Much of what follows gives detail on that achievement.

Let us just recap on the conditions necessary for that achievement (of course, it is not possible to be definitive on what exactly was most necessary, or what might have perhaps been less essential):

- skills and commitments of professionals;
- skills and commitments of the people involved "at the base" in neighbourhoods;
- energy spread even more widely than this, released after the end of a dictatorship — Francoism;
- ideology contained within a broadly hegemonic political project;
- wealth in the city built up over long periods;
- other historical-geographical features giving some "winning cards";
- the global historical moment of the 1970/80s.

To judge whether the more "structural" or "agency" features of these conditions were more important would be to enter the most contested zones of history and the social sciences. Was the opportunity really quite specific to those years, or is there something more fundamental in the Barcelona/Catalan air which has a longer drive? No doubt we must try to keep our focus simultaneously on a range of factors. The deep structures facilitated by the contemporary movements of capitalism and state forms have mattered a very great deal in Barcelona: just compare the difference with the equivalent structuring forces affecting the trajectories of say London, Rome or New York to start to get a feel for the power of these active mechanisms. Equally, the agents, individually and as wider interests, fought and struggled to achieve what has been achieved; it would not have happened in the same way at all otherwise. The core actors had social democratic or socialist ideals. Barcelona's success is in part a success story of a certain kind of Left.

At any rate, the whole period was still deeply contested, between more egalitarian and more conservative forces, so that however much one may see a broad hegemony, this was never a settled or fundamental consensus. Now, around the achievements of the recent past and the plans put together since about 1998, there is again less consensus in the city, at least amongst the interested experts. Almost certainly the level of technical skills remains high. Schemes, such as that for the 2004 Forum, will very likely be delivered on time and to an acceptable standard. The Besòs river area will probably be greatly improved in the next few years, and the Llobregat projects will be completed — port, airport, transport links — if not all to the satisfaction of environmentalists. Larger question marks may hang over the 22@ scheme, but this also is highly innovative. Whether the intensive mix of residential and business uses will emerge and work in this vital new central district will only be known over a 10–20 year period.

That scheme and others will be dependent largely on a development model which is increasingly generalized in Europe, whereby private profitable schemes pay for public gains, simply because public budgets are seen as not able (or willing) to carry the costs. That is the fundamental issue underlying the prospects for the city's planning. Now that the coalition of progressive forces which impelled the earlier project has, arguably, more or less lost its ideological strength, does the development model inevitably slide into the more normal pattern, normal often in Barcelona in past eras and normal in most of Europe now?

The challenges to such an assumption of "return to normality" are various, and could be connected to each of the conditioning factors listed above. One is the question of the economic solidity of the city and region. This now depends on the three legs of manufacturing within a continental division of labour, advanced services partly serving that manufacturing and partly working more widely, and tourism. Economic success has been considerable since the mid-1980s, on these bases, and could plausibly continue. However, these depend on many wider economic and environmental conditions which could change dramatically; cheap energy of various forms is just one, and critical for all three legs.

Social solidity is also an issue. The dimension of this most commented on since about 1990 has been the challenge of immigration from poorer countries, especially northern Africa. Barcelona has become, like most other European cities, a multinational city, along several dimensions. This cheap labour is essential to the effective functioning of the city, especially to the tourist trade, but causes some social conflict, and always threatens to emerge over the political parapet and drive politics towards less tolerant and liberal practices. Parts of the city and region have become more segregated by ethnic group. This becomes an issue in planning, as it did in other countries with large immigrations much earlier, but it is not an issue which has yet been much addressed.

The nature of political agency is perhaps the biggest question mark. Since 1980 the city and Catalonia as a whole has had a relatively stable political landscape. Although there have been undoubted drifts in emphasis, a broad commitment to public action for wide sectors of the citizenry has continued in public discourse. This has been a vital accompaniment to planning, from the level of the design of quality street furniture to the debates on the location of the high-speed train station (in la Sagrera, an area of regeneration). Any shift towards a more US style model that puts private needs and dynamics in the centre of programmes would clearly challenge this functioning tandem of action. Given that such changes have happened to varying degrees in most other Spanish cities, such a panorama is by no means implausible. Equally though, pressures from other directions — sharpening housing problems, environmental stresses, disenfranchised immigrant groupings, rising unemployed — could drive towards the creation of a quite fresh coagulation of political interests. Whatever the direction taken, the action will play out in a city which was comprehensively transformed in the last quarter of the twentieth century, in one of the most sustained bursts of planning and conscious governance seen anywhere at the urban level. Probably that burst will not be forgotten for a long while. It may then generate its own legacy.

References

Associació Pla Estratègic Barcelona 2000 (2000) 1*0 Anys de Planificació Estratègica a Barcelona 1988–1998*, Barcelona: Associació Pla Estratègic Barcelona 2000.
Bohigas, O. (1985) *Reconstrucciò de Barcelona*, Barcelona: Edicions 62.
Borja, J. (2000) *Ciudad y Ciudadania: Dos Notas*, Barcelona: Universitat Autonoma de Barcelona, Institut de Ciencies Politiques i Socials, Working Paper No. 177.
Borja, J. and Castells, M. (1997) *Local and Global*, London: Earthscan.
Brunet, F. (1995) "An economic analysis of the Barcelona '92 Olympic Games: resources, financing and impact", in M. de Moragas and M. Botella (eds), *The Keys to Success: the Social, Sporting, Economic and Communications Impact of Barcelona '92*. Barcelona: Centre d'Estudis Olimpics i de l'Esport/Universitat Autonoma de Barcelona, pp. 203–37.
Busquets, J. (1992) *Barcelona*, Barcelona: Fundacion Mapfre.
Carreras, C. (1993) *Geografia Urbana de Barcelona*, Barcelona: oikos-tau.
FAVB (Federaciò d'Associacions de Veins de Barcelona) (1999) *La Barcelona dels Barris*, Barcelona: FAVB.
Huertas, J. M. and Andreu, M. (1996) *Barcelona en Lluita*, Barcelona: FAVB.
Lauria, M. (ed.) (1997) *Reconstructing Urban Regime Theory*, London: Sage.
McNeill, D. (1999) *Urban Change and the European Left. Tales from the New Barcelona*. London: Routledge.
McNeill, D. (2002) "Barcelona unbound: urban identity 1992–2002", *Arizona Journal of Hispanic Cultural Studies*, 6, pp. 245–61.
Magrinyà, F. and Tarragó, S. (1996) *Cerdà: Urbs i Territori/Planning Beyond the Urban*, Madrid: Electa.

Mancomunitat de Municipis (1996) *Dinàmiques Metropolitanes a l'Àrea i la Regió de Barcelona*, Àrea Metropolitana de Barcelona.

Marshall, T. (1994) "Barcelona and the Delta: metropolitan infrastructure planning and socio-ecological projects", *Journal of Environmental Planning and Management*, 37(4): 395–414.

Marshall, T. (1996) "Barcelona — Fast Forward? City Entrepreneurialism in the 1980s and 1990s", *European Planning Studies*, 4(2): 147–65.

Marshall, T. (2000) "Urban planning and governance. Is there a Barcelona model?", *International Planning Studies*, 5(3): 299–313.

Moix, L. (1994) *La Ciudad de los Arquitectos*, Barcelona: Anagrama.

Moreno, E. and Vázquez Montalban, M. (1991) *Barcelona, Cap a On Va?* Barcelona: Llibres de l'Index.

Nello, O. (2001) *Ciutat de Ciutats*, Barcelona: Empúries.

Narotzky, V. (2000) "'A different and new refinement'. Design in Barcelona 1960–1990," *Journal of Design History*, 13(3): 227–43.

Raventós, F. (2000) *La Col·laboració Publicoprivada*, Barcelona: Aula Barcelona (also available in English at www.aulabcn.com).

Riera, P. (1993) *Rentabilidad Social de las Infraestructuras: las Rondas de Barcelona*, Madrid: Civitas.

Salmon, K. (1995) *The Modern Spanish Economy*, Pinter: London.

Salmon, K. (2001) "The Spanish economy: From the Single Market to EMU", in R. Gillespie and R. Youngs (eds) *Spain: The European and International Challenge*, Frank Cass: London.

Santacana, F. (2000) *El Planejament Estratègic*, Barcelona: Aula Barcelona (also available in English at www.aulabcn.com).

Sharpe, L. J. (ed.) (1995) *The Government of World Cities*, Chichester: Wiley.

Soria y Puig, A. (1995) "Ildefonso Cerdà's general theory of urbanizacion", *Town Planning Review*, 66(1): 15–40.

Trullen, J. (2001) *La metròpoli de Barcelona cap a l'economia del coneixement: diagnosi econòmica i territorial de Barcelona*, Barcelona: Ajuntament de Barcelona.

Vilar, P. (1964) *Catalunya dins l'Espanya Moderna*, Barcelona: Edicions 62.

Ward, S. (2002) *Planning the Twentieth Century City*, Chichester: Wiley.

PART 1

What happened

In the space of eight chapters, it is only possible to give an overall picture and a few flavours of what happened in Barcelona in the 1980s and 1990s. The overall picture comes at any rate from some of those most closely involved, including the mayor Pasqual Maragall, another leading councillor in a whole range of planning-related initiatives Jordi Borja, the council's director of planning in the crucial early years Oriol Bohigas, and two professional planners who, in their work in the city council, were deeply involved in both framing policies and implementing them, Juli Esteban and Amador Ferrer.

The other writers were generally less involved, and write from varying academic situations; one, Nico Calavita, outside the country, but a frequent visitor, the other two, Nuria Benach and Oriol Nello, within Barcelona's universities, although Oriol Nello has been increasingly involved in the field of planning and metropolitan politics over recent years, and since December 2003 is a senior member of Pasqual Maragall's Socialist team in the Catalan government.

Together these key actors not only convey information, but are giving testimony of their perspectives on the development of events. Clearly, this part thus lays no claim, any more than the rest of the book, to being an objective account of what happened, whatever that may look like from some more remote historical standpoint.

The chapters deal with quite varying slices of what has happened in urban change and the role of planning in the city. Nello's takes a wider view, looking at Barcelona within its metropolitan region, whilst that of Calavita and Ferrer gives a more historical account of how the residents movement of the 1970s crucially shaped so much of what followed.

Maragall's chapter, originally published in 1997, looks back on his mayoralty, ranging far beyond planning matters, in the form of a public letter addressed to the presidents of the governments of Catalonia and Spain. Its special contribution is to show how planning fits into a much wider political model and programme of city governing, conveying some of the thought processes of a politician of perhaps rare intellectual abilities — as well as being an effective election winner. The chapter may be hard for those unfamiliar with Barcelona. The editorial decision not to decorate the chapter with footnotes will, it is hoped, be preferred, but some detailed allusions will understandably have to be jumped over by many readers.

Another chapter with a political framing role is that by Borja on governance, showing again how important this element was in the Socialist programme. The first 12 years or so can be seen as a protracted campaign to change the nature of urban governing, towards a more citizen-based democracy. Borja's schooling in Paris in the 1960s and in the residents movement in Barcelona, as well as his academic engagement as an urban sociologist, provided him with much of the foundation and force for the years he spent leading key council initiatives.

Two chapters are more "professional" contributions, by Bohigas and Esteban. One gives a brief taste of the perspective of the "grand old man" of Barcelona architecture, who has been designing buildings and writing of Catalan architectural history since the 1950s. It is to be hoped that more of Bohigas' acerbic commentary on Barcelona planning will be translated into English in the near future (his latest role as a Catalanist and controversial newspaper columnist resulted in a book, *Cartes de la Baralla*, Columna, 2002, which is less likely to have an international audience). Esteban, by contrast, gives a sober and carefully judged account of the steps involved in changing the city, as seen in the mid-1990s, in part drawing on his years running courses on planning in the architecture school of Barcelona. A strong feature of the Barcelona experience is the closeness of academia and practice in the city — virtually all these actors have, either simultaneously or before and after, also been teaching the next generation of professionals in university postgraduate courses.

The final chapters in this part are on a much praised aspect of the transformation, public spaces. Benach gives a more distanced and analytical treatment, a chapter taken from a book on public spaces written mostly by Barcelona University academics (*Espais Públics: Mirades Multidisciplinàries* edited by Rosa Tello i Robira, published by Portic in 2002). The collection of short pieces introduced by Borja is selected from far more contributions in a second book published on the topic, looking at new public spaces all over the world in recent years (*L'Espai Públic: Ciutat i Ciutadania*, edited by Jordi Borja and Zaida Muxí, and published by the Diputació de Barcelona in 2001). These two selections amongst a growing local documentation of the topic are enough to give some idea of how the experience is being reviewed, reflected upon, perhaps advanced or changed, and certainly diffused globally as one thread of the city's best practice.

Chapter 2

Urban dynamics, public policies and governance in the metropolitan region of Barcelona

Oriol Nello

Introduction

For the past two centuries Barcelona has essentially been an industrial city. Spain's first steam powered industry was established here, in 1832, its first railway line was built here, in 1848, its leading textile centre ended up being formed here, as well as its most active port, and its most populous working class concentration. As a result, for many years the city's image was directly related to industrial activity as well as to the social conflicts that its development entailed: "Spain's factory", the "city of bombs", the "rose of fire".

Had we, during the 1970s, been able to explain to a resident of Poblenou (the most industrial neighbourhood of the most industrial city, the "Catalan Manchester", cradle of the proletariat and anarchist movement) that, in a mere 30 years, the biggest factories (Titan, Motor Ibérica, la Maquinista Terrestre y Marítima) would completely disappear and that it would flourish into a tourist, residential and cultural district, that resident would have had a very hard time believing it.

And yet, this is exactly what has happened. In the last quarter of the twentieth century, Barcelona has undergone a radical transformation: its economic base, its social structure, the population's habits, its physical structure and even its image have experienced accelerated changes that have been decisive and, in general terms, positive. Urban policies have had an outstanding role in this evolution, whose results have been so spectacular that they have even attracted international attention.

The object of this chapter is precisely to provide some data and observations on the nature of these changes and their relationship to the public policies that have been applied in the city. Thus, it will try to develop the five following propositions.

a   Barcelona now constitutes a metropolitan reality subject to intense transformation dynamics. These cause the city to integrate an ever-growing space, to see its activities and population disperse, as well as to functionally and socially differentiate the different areas that integrate them.
b   This transformation, which has been accompanied by a structural change in the economic base and a sharp increase of average income levels, offers manifest opportunities, amongst which we can highlight those which are derived from the jump in metropolitan scale and from improved access to services and to jobs for the whole urban area.
c   However, these contingencies are also accompanied by important challenges that can put environmental sustainability, functional efficacy and the city's social cohesion at risk.
d   With regard to these opportunities and risks, there has been an attempt to develop a set of policies that tend towards safeguarding competitiveness, functionality and the central city's quality of life, by defending a model of urbanization characterized by density, complexity and social cohesion.
e   The reasonable success achieved by these policies, in which there are, to be sure, both lights and shadows, makes the case of Barcelona interesting for a general reflection on urban policies and governance in large contemporary cities.

The dynamics of urban transformation

We have stated that the metropolitan area of Barcelona is currently going through an accelerated process of change. There are three characteristic tendencies of this transformation, which, as we will see, match closely the dynamics of transformation of most of the big cities in the Iberian Peninsula and in Western Europe.[1]

a   *Dispersion*: both population and activities, after a long process characterized by an accentuated tendency towards concentration, now disperse across the metropolitan area.
b   *Extension*: at the same time as urban sprawl across the territory, there tends to be a progressive expansion of this area in order to integrate an increasingly large area within the metropolitan boundaries.

c   *Specialization*: this dispersed and expanding city also tends towards functional and social specialization of each one of its areas, in relation to the whole.

We will explore how these dynamics work in greater detail. But before we go any further, a brief reminder of the magnitudes that we are discussing is convenient.

The metropolitan region of Barcelona: basic magnitudes

The basic data that refer to space, economic activity and population in metropolitan Barcelona are the following.[2]

The metropolitan area of Barcelona, in its administrative definition, is of 3,235 km², which represents slightly over 10% of the territory of the Catalan region.

The metropolitan area's GNP, with a gross amount of 86.4 million euros, represents 69% of total Catalan GNP and 13.4% of total Spanish GNP (the Catalan GNP is 125.4 million euros and the Spanish, 646.8 million euros, according to estimates available for the year 2001). Likewise, we must point out that the province of Barcelona absorbs 25.6% of Spanish imports and generates 22.3% of exports.

- The total population of metropolitan Barcelona, which grew incessantly and very visibly in the 1960s and the beginning of the 1970s, has remained practically stable since 1981. In 1996, according to data from the last available census, it reached 4,228,047 inhabitants. This represents 69.4% of the Catalan population.
- This population is very unequally spread out, so that in the hardly 100 km² of the municipality of Barcelona there are 1.5 million residents, in about 30 surrounding municipalities (over a surface of 378 km²), another 1.3 million, and, finally, in the remaining 2,759 km² of the metropolitan territory there is another 1.4 million. Thus, one-quarter of the Catalan population finds itself in 1% of the territory of Catalonia, almost half in 2%, and approximately three-quarters in 10%.
- Finally, we must point out that the metropolitan area's urban structure is rich and complex since, along with the central city, it includes an ensemble of medium-sized cities (of between 30,000 and 200,000 inhabitants) some of which have an important industrial and commercial tradition (like Mataró, Granollers, Sabadell, Terrassa, Vilanova or Vilafranca) whereas others are asserting themselves as emerging residential and tertiary centres (like Sant Cugat, Mollet, Cerdanyola, Sitges or Calella).

Sprawl of population and economic activities
Currently, the first of the dominating tendencies in the configuration of the metropolitan area is the dispersion of urbanization over space.

In fact, the spatial structure of the metropolitan region as we know it today is the result of a long process of concentration of both the population and its activities within the Catalan region. This is a process which goes back to at least the eighteenth century and through which, throughout the agricultural and commercial revolution of the 1700s, the industrialization of the 1800s and the modernization of the first three-quarters of the 1900s, the population, coming from remote areas of Catalonia and even from other regions of Spain, tended to concentrate along the coastline and, specifically, in the plain of Barcelona.

This process of concentration resulted in both the inequalities in the distribution of the population to which we referred above and one of the most noticeable characteristics of the city of Barcelona: its extremely high density (15,000 inhabitants per km²), for which it is hard to find parallels in other European cities.[3]

This process of concentration reached its zenith in 1981 when Barcelona attained its highest demographic density in history (1,752,627 inhabitants). Since then it has known a certain reversal, a decentralizing and dispersing ripple that, in an irregular yet continuous manner, has been affecting the whole of the metropolitan area. Thus, in the past twenty years, the city of Barcelona has lost close to 250,000 inhabitants and has gone from containing 40% of the metropolitan population to 35%. Meanwhile, the first metropolitan ring remains stable from a demographic point of view (both in absolute as well as in relative terms) and the second ring grows at an accelerated rate (Figure 2.1).[4]

2.1
Parc Tecnologic del Valles in the late 1980s, showing part of the district with intensive development since that date in the wider metropolitan region

We thus find ourselves facing a real decentralizing phenomenon, with net losses of population from the metropolitan centre that affect not only Barcelona but also the municipalities surrounding it (like l'Hospitalet, Badalona or Santa Coloma, each of which experiences population loss). This is a phenomenon that, on the other hand, is in no way original, since, as we know, it affects a large part of the Spanish metropolitan areas (thus, of the seven largest Spanish cities — Madrid, Barcelona, Valencia, Bilbao, Sevilla, Málaga and Zaragoza — the first four have experienced net population losses in their central municipality in the last inter-census period).[5]

The decentralizing process is the result of important inter-metropolitan residential migrations. The detailed analysis of these allows us to draw two consequences:

a   The configuration of the metropolitan space has passed from being narrowly conditioned by inter-regional migrations associated with the labour market (in the phase before 1975) to depending, above all, on inter-metropolitan migrations associated with the housing market.

b   The phenomenon we are facing is not simple decentralization, but rather a real process of dispersion of the population and activities across space. This is a process through which practically all the nuclei with the highest populations and densities (independently of their locations) tend to yield relative weight and, in many cases, population in absolute terms, towards other more dispersed and less densely populated locations.

Demographic dispersion also corresponds to the growing dispersion of economic activities and services in the metropolitan territory. In the past 20 years, the city's economy has lived through a very pronounced structural adjustment: the shift from an industrially based economy to an increasingly tertiary one, as well as expanded flexibility in the productive process (which has affected both the average size of businesses, and the organization of productive processes and of labour regulations). Thus, this double process of adjustment has been complemented by, from the spatial point of view, a noticeable tendency towards decentralization of population and jobs over the area.[6]

Available data show that Barcelona still maintains a significant relative weight in total employment (in 1996 it still retained 659,786 of the 1,525,090 jobs in the metropolitan region), while it loses positions rapidly: between 1975 and 1996, the city went from holding 56.2% of the jobs in the metropolitan region of Barcelona to 43.3% (while the first ring remained stable in relative terms — 23.7% in 1975 and 23.5% in 1996 — and the second grew in an accentuated manner — from 20.1% to 33.2%). The latest available data from the Metropolitan Survey for the year 2000 suggest that the tendency continued in the past five years, in such a way that the weight of Barcelona today would hardly surpass 40% while the second ring would already reach more than 35%.[7]

Urban area expansion

The second contingency that now characterizes the evolution of this metropolis is its expansion over space. At the same time that its population and its activities disperse over the territory, the metropolitan region expands more and more in order to integrate an ever-growing space in its network of daily functions.

Thus, if we try to define the scope of metropolitan Barcelona in terms of conventional criteria of the SMSA (Standard Metropolitan Statistical Area) we see how the territory that we can consider metropolitan has gone from:

- 62 municipalities and 1,010 km² in 1981; to
- 146 municipalities in 1991; and to
- 216 municipalities and 4,597 km² in 1996.

In this way, the strictly metropolitan area surpasses its own administrative boundaries in terms of planning (163 municipalities) and penetrates strongly towards the interior of Catalonia and the provinces of Girona and Tarragona.[8] In the immediate future we must observe attentively the effects that the introduction of the high-speed train between Barcelona, Girona, Lleida and Tarragona (cities situated at a radius of 100–150 km from Barcelona) will have on these expansive dynamics.

Functional and social specialization

Finally, these tendencies of dispersion and expansion of metropolitan dynamics are further accompanied by a third phenomenon: the growing specialization of each one of its components (of each municipality, each neighbourhood) in relation to the metropolitan whole.

Thus, even if it is true that demographic, occupational and service dispersion over the metropolitan territory has tended to reduce differences between the centre and the metropolitan rings (even in terms of income, as we will see below), the growing integration has brought the specialization of each municipality in functional terms. In this way, residence, commerce, industry and leisure activities tend to become increasingly diverse within the metropolitan space.

This specialization is also followed by a greater segregation in social terms. A moment ago we were referring to metropolitan migrations, migrations that, as we said, directly affect the housing market. That is, the dominant flows of residential mobility are induced and regulated, to a great extent, by the capacity of individuals and families to act in the land and housing market. This market displays very important rigidities.

a   According to data from the Metropolitan Survey 2000, 84.6% of the metropolitan population is made up of families that own the dwellings they are living in, which leaves the rental market in a very secondary position.

b   On the other hand, the stock and production of officially protected housing is very small in comparison to other European countries. Thus, only 1 of every 25 units of housing produced in the year 2000 in Catalonia had some form of public protection.[9]

The economic effort necessary for the acquisition of housing is very high (up to 7.8 times the annual disposable family income in the case of new housing in Barcelona), and it has increased noticeably in the last few years in spite of the reduction of interest rates. In this context, the difference in price between the central city and the metropolitan rings (in the city prices are 1.4 times higher than in the first ring and 1.8 higher than in the second) works as a strong stimulus for the selection and departure of those who can afford housing in the metropolitan rings but not in the central city (that is, sectors of population that are mostly young and whose levels of education and income are above the average). Price differences inside each one of those rings act in the same way in the regulation of population movements.[10]

We thus see that, in this context of market rigidity, of high prices and few public policies, the housing market acts as a powerful filter that conditions the process of urban segregation.

The benefits of metropolitanization

We therefore see that in Barcelona, as in many other European metropolises, current economic and social transformations entail a process of metropolitanization that is radically altering the form, the function and the cohesion of urban spaces.

We should point out that this transformation can lead to some striking positive consequences, of which we may highlight the following.

a   *Increase of average income*. The process of metropolitanization (with the unification of the labour market, the structural change of the economy and the emergence of new sectors) has been both cause and consequence of the economic growth experienced in Catalonia and in Spain in recent years. Economic growth has been faster, in average terms, than in the rest of the European Union, which has permitted Catalonia's per capita income, which in the year 1986 was still 86% of the Union's, to reach 100% in the year 2000.

b   *The metropolitan scale leap*. On the other hand, the city's transformation from 2.5 to 4.3 million inhabitants effectively integrated in the labour and the metropolitan consumer markets has given Barcelona a critical mass which places it in a much better position to attract private investments, services, business headquarters and investments in public works. Thus, Barcelona constitutes the sixth metropolitan region of Europe, only surpassed by London, Paris, the Ruhr, the Randstad and Madrid.[11]

c   *The relative balance of income levels.* There has also been a certain convergence among income levels in the city and in the metropolitan rings. Thus, the average income levels of the metropolitan rings, which used to be much lower than those of the central city, are now closer to these. In this way, the average income level of the first metropolitan ring that, in 1989, was equivalent to 77.1% of the central city's, had already reached 85.3% in 1999. And the second ring, in the same period, went from 79.8% to 96.8% with respect to the average income level of Barcelona.[12]

The challenges of urban transformation

However, it is undeniable that these advantages have been accompanied by important problems. Thus today, the city, and the metropolitan area as a whole, must face challenges of an environmental, functional and social order.

The environmental problem

In terms of the environment, the most outstanding problem is, without doubt, land consumption. This has undergone an exponential increase, from the situation in 1882, when only 1,763 of the 323,000 hectares of the metropolitan region of Barcelona were occupied by urbanization, to 21,482 hectares in 1972 and 45,036 hectares in 1992.[13] Currently, the rhythm of transformation is close to 1,000 hectares per year, that is, approximately 3 hectares per day. The result, as one can see, is quite striking, and it implies:

- a liability for future development and the public interest;
- the isolation and sacrificing of spaces of natural interest;
- occupation of land of outstanding agricultural value;
- loss of landscape values;
- impermeabilization and artificial landscaping of large amounts of land, with the corresponding increase of natural risks.

Furthermore, urban sprawl translates into increased energy consumption, greater difficulty in gathering and treating waste, as well as a higher rate of water consumption per inhabitant, and health problems. Finally, as planning literature has been asserting for years, sprawl also noticeably affects mobility.

Functional risks

Expansion of the metropolitan area and dispersion of urbanization have entailed an explosion in the mobility needs of citizens and businesses. In this expanding urban area, in which every place is specialized in reference to the whole, citizens now use space in a more extensive way. They therefore carry out functions (residence, work, shopping, enjoying leisure time), which used to be restricted to a reduced area, over an increasingly vast space.

This is very visible in the evolution of the municipalities' degree of autonomy (that is, their capacity of retaining the mobility that is generated within their own nucleus).[14] Thus, in the province of Barcelona, municipalities containing less than 50% of their labour mobility have passed from:

- 102 of the 310 municipalities in 1986, to
- 151 of the 310 municipalities in 1991, to
- 208 of the 310 municipalities in 1996.

In this way, in 4 out of every 5 municipalities of the metropolitan region of Barcelona, at least half of those who work do so outside of their own municipality and the average rate of autonomy dropped from 67.6% to 55% between 1986 and 1996. Data from the Metropolitan Survey of the year 2000 show a new drop of 7 percentage points between 1995 and 2000.[15]

This expansion in mobility has entailed an extraordinary increase in the demand for road infrastructures and for public transport. Nevertheless, since investment in this last field is much lower than that of the road network, the modal split has suffered a radical modification in favour of journeys taken in private vehicles. In the city of Barcelona alone and in the displacements associated with it, public transport still retains a certain strength and it represents just over one-third of journeys taken (Figure 2.2).

This situation is, without a doubt, the result of investment policies (public and private) that have given priority to investments in the road

2.2
Traffic volumes in metropolitan area 1994, showing importance of new city ring, and of motorway box inland from the city

network to the detriment of public transport. But it is also the consequence of the way land occupation has taken place in recent years. Thus, according to data from the agency Barcelona Regional:[16]

a   only 2% of the surface area of the metropolitan region has average accessibility to the rest of the metropolitan territory equal to or lower than 60 minutes;
b   while 44% of compact (higher density) residential land is within the area of direct influence (500 m) of a station;
c   only 6% of dispersed residential land and 11% of industrial land enjoy this condition.

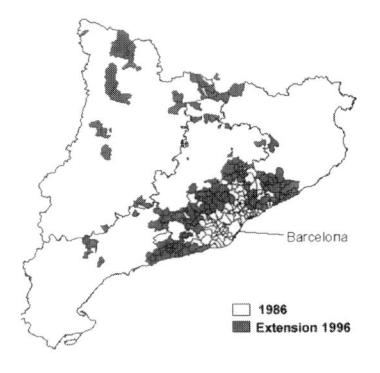

Thus, in the past few years, sprawl has entailed that a higher percentage of the population and economic activities are now established in low accessibility areas (Figure 2.3).

The new patterns in land occupation, along with scarce investment in public transport, are leading to problems in the metropolitan mobility networks, expressed in:

- congestion of the road network;
- increased commuting time (especially affecting low income groups and, within each household, women, children and the elderly);
- growth of pollution and accidents;
- increased energy consumption per worker and per place of work.

2.3
Change of scale of the metropolitan region between 1986 and 1996, from data showing intensity of linkages to central zones

Social segregation
We have explained how the housing market acts as a powerful incentive for the separation of groups over metropolitan space, in particular, for the emigration of medium income groups from the central city towards the metropolitan rings. These phenomena could lead to dualization of the city, in the sense that it could become a place suitable only for the very rich or for the very poor.

Statistical evidence,[17] fortunately, denies this hypothesis.

a   As we have explained above, differences between the central city and the metropolitan rings, in terms of average incomes, have tended to diminish in the past 15 years.
b   But, furthermore, in the central city, the distance between the first and the last decile of the population in terms of average income has decreased from 15 to 1 in 1985 to 10.5 to 1 in the year 2000.
c   Similarly, total income distribution both in the central city and in the metropolitan rings has tended to become more equitable (as the analysis based on the calculation of Gini indexes shows).

This positive evolution of the differences in income has to be attributed, in good measure, to the introduction of the basic mechanisms of the welfare state in Spain (non-contributive pensions, unemployment coverage, universalization of health care) and to urban policies which we will address later.

In any case, these advances do not in any way deny either the persistence of important inequalities (the first quintile of the population obtains 40% of total incomes, while the last only receives 7.3%), nor the continuation, and even the aggravation, of very problematic situations for certain groups and certain areas. Among these, because of the severity of the problems that affect them, we can highlight a number of metropolitan neighbourhoods that represent 6% of total residential surface area, 180,000 homes and close to 400,000 inhabitants that, according to the administration, should be the object of "integral rehabilitation". We refer to, especially, deteriorated central areas and housing estates built in the 1960s and 1970s, as well as some areas of marginal urbanization. This is where the sectors with the lowest acquisitive power and the new non-European immigrants tend to concentrate nowadays.

Urban policies and the question of governance

We have tried to show how the metropolitan region of Barcelona currently finds itself in an accelerated process of change. The city, which has traditionally been characterized by its physical compactness, by the complexity of uses and by the coexistence of very diverse social groups, is becoming more dispersed, more specialized and potentially more segregated.

As we have shown, this transformation entails opportunities and challenges for the future of the central city and its area. The challenges are particularly important for the central city which, left to itself (that is, if there were no regulation of any sort over the impulses of the market), could become an increasingly inefficient space in functional terms, unsustainable in environmental terms and inequitable in social terms.

The urban policies that municipal governments have been trying to implement in the past two decades have, in great measure, had to face these challenges as an essential objective. In order to do so, they have defended the model of a compact, complex and integrated city. That is, we understand compactness, complexity and cohesion as essential prerequisites for efficacy, sustainability and equality: the unrestrained sprawl, the dispersed city, is not a city, at most it is urbanization, in which those elements that make contemporary urban life attractive and desirable (innovation, richness of uses, capacity to compensate for inequalities) are anything but present.

In this vindication of the city, urban renovation has played an essential role. And it has achieved considerable successes in Barcelona, such as the modernization of infrastructures that took place for the Olympic Games in 1992, the rehabilitation of Ciutat Vella, the intervention in Poblenou or the public spaces policy in all of the districts, for both the centre and periphery. Facing the future, the new policies of renovation and defence of the central city will have, in our opinion, to face the challenge of creating both a vision and the appropriate instruments, to tackle four basic questions.[18]

a   *Integration in the network.* In a world of constant fluxes, the city must be a place that is sufficiently attractive to exercise the function of a hinge between the logic of a global order and that of local order. This implies assuring its accessibility to the exterior, its capacity for innovation, as well as the diversity and coexistence of people of heterogeneous origins and expectations in its interior. Access to the exterior, innovative capacity and internal diversity are the keys for founding the city firmly as a node of the global urban system.

b   *The articulation of the metropolis.* The compact city (the intense city as we have sometimes referred to it) is in a better position than others to face these challenges. But we would be mistaken if, when defending the compact city, we were to deny metropolitanization or to try to base metropolitan development on the potential of only one urban centre. If we did so, the pressure on the central city would be so intense that it would translate, through prices, into new segregation and exclusion processes. What we must pursue is a metropolis articulated by a number of cities integrated into a network, a polynodal metropolis, a city made up of cities: a city of cities.

c   *Innovation with roots.* The type of urban renovation that derives from these postulates must be able to combine transformation with memory, creation with inheritance, progress with equity. And all of this, in the central city, requires urban policies that:
   – establish cultural projects as a motor and as irradiation;
   – defend and strengthen public spaces;
   – search for the connection between the historical substratum and new types of activities;
   – evade untamed population movements and the formation of ghettos;
   – give priority to rehabilitation over demolition;
   – link land uses to accessibility;
   – regulate traffic.

   In a nutshell: the future of the central city nowadays depends, on the one hand, on its capacity to renovate itself through the values of coexistence and innovation, and, on the other, on its capacity to lead a region with structured space and of high environmental quality. This capacity is directly related to the city's ability to create, stimulate and maintain activities in innovative and emerging sectors.

d   *The centrality of politics.* In order to advance towards this urban configuration we must have values and collective projects capable of conditioning and contradicting, whenever it is necessary, market impulses. We must therefore claim the central role of politics in the construction of urban space. To navigate along with the flow is not to govern. To govern means to make something happen that without decision and collective will would not occur. This does not mean, however, that we should be slaves to old plans, to fixed schemes. Any urban policy that remains fixed quickly becomes obsolete. Thus, the defence and stimulation of urban renovation must imply:
- flexible planning that is more engaged in the process of construction of the city than of following a final image-objective;
- administrative and strategic planning that is capable of considering the decisive options for economic and social transformation, along with the physical content;
- participatory planning, which includes electronic interaction, simulation exercises and citizen advice;
- planning, in short, which inscribes sustainability, efficacy and equity as essential objectives.

The city of quality is not one which intends to annul its conflicts, and reach quality in such an unreal way, but rather one which is able to administer these contradictions through democratic processes and collective projects.

The role of emerging sectors: three examples

Finally, we will examine three examples of the way in which urban renovation policies trying to put these values into practice are being applied in the city of Barcelona. We will explore three examples that, because of their magnitude, have implications for all aspects of the lives of citizens. However, because of their characteristics, each one of them has, respectively, a special impact on compactness, complexity of uses and the city's social cohesion, i.e. each one of the values they are meant to preserve.

Tourist and cultural activity, and the new leisure centres

Tourist activity is a sector that clearly demonstrates expansion in the city of Barcelona. Although it is true that Catalonia, particularly the coastal areas of the Costa Brava and the Costa Dorada, constitutes one of the main tourist destinations of the Mediterranean, the city had remained relatively outside of major tourist flows. It was mostly visited for business reasons, and in particular for its role in conference and trade fair organization, which is one of its traditional occupations.

This situation has experienced a radical change since the celebration of the Olympic Games in 1992. These had, from the point of view of stimulating tourist activity, a double virtue:

a   on the one hand, they constituted a colossal exercise of promotion for the city, that allowed it to project an image based primarily on its cultural and architectural charm;
b   on the other hand, attracting public and private investment for the city, with very noticeable effects in the area of infrastructures and facilities, also included the carrying out of a "Hotel Plan". This allowed the city to move from 118 establishments and 18,569 beds in 1990 to 148 establishments and 25,055 beds in 1992. Since then, the supply has not stopped growing and in the year 2000 it reached the figure of 187 establishments and 31,338 beds. Up to the year 2004, the addition of 7,500 new beds is foreseen.

Simultaneously, demand has also experienced a very sharp increase, which has led it from 3.8 million overnight stays in 1990 to more than 7.7 million in 2000, doubling in one decade. Some 29% of the visitors are Spanish, and, of the rest, those who come from the following countries stand out: the United States (13.5%), the UK (11.2%), Italy (6.3%) and Germany (5.8%).

Another element that has experienced a noticeable change is the reasons for these visits. While at the beginning of the 1990s, business visits amply surpassed 50% of total visits, in the year 2000 these only constituted 36.7%, while tourism already reached 43.9%. Furthermore, incoming tourism is one with high cultural interest that, according to surveys carried out, values as the city's three principal attractions its provision of impressive architecture, culture and entertainment.[19]

This increase in external demand for culture is complemented by a noticeable increase in internal demand. A substantial part of the population of the metropolitan region of Barcelona considers that it has a lot of free time (29.1%) or quite a large amount of it (34.5%; although this leisure time is often attained by taking time from sleep: in the city of Barcelona, in the year 2000, 60.2% of the population stated that they go to sleep between midnight and one in the morning, while 71.6% stated that they wake up between 5 and 8 a.m.). Amongst the adult population, leisure time activities that are "frequently" practiced are: going to the beach (42.2%), to the movies (19.9%), to restaurants (16%) and to museums and exhibitions (8.6%).[20]

Urban policies have tried to associate the emergence of this new demand with the creation of new centres for leisure time activities within the existing city. These centres have also been used as instruments of rehabilitation in some areas and invigoration of others. On the whole, the city has bet on tourism, cultural and urban leisure activities, in contrast with the dispersion of urbanization, land consumption and the banalization of the landscape that the tourist development of the coast has entailed.

The most important of these new centres of cultural and leisure time consumption are the following.

a   *The Ciutat Vella cultural axis.* This crosses the Raval, in the middle of a historic part of the city, from the port to the Placa de la Universitat. Here, in the heart of a neighbourhood with intense deterioration problems, some of the main cultural facilities of the city and of Catalonia have been placed: the Museum of Contemporary Art of Barcelona, the Centre of Contemporary Culture of Barcelona, the Centre of Cultural Resources, headquarters to one of the foremost Catalonian editorial groups, the renovated National Library and the headquarters of the Higher Committee on Scientific Research.

b   *The waterfront.* From Barceloneta to Poblenou and La Mar Bella close to 5 km of urban coastline have been recovered, and these have been adapted for swimming and leisure activities, with the construction of a seaside promenade, the Olympic Port, and the establishment of numerous restaurants and bars. This opening of the city to the sea, which must, without a doubt, be considered as one of the main legacies of Olympic planning, has allowed what the city's tourism can offer to change radically (sun and beach as new products), along with the leisure habits of its inhabitants (the sea front has turned into many people's favourite beach: 27.2% of people who go to the beach state this as their main destination).[21]

c   *The Port Vell.* The inner harbour of Barcelona, adjacent to the city's historic centre, has also been recovered for citizen use, with the removal of the barriers that used to prevent access to the sea and the construction of walkways that allow the old quays to be reached. Here, the Maremagnum centre has been established, fully dedicated to leisure time activities, with a shopping centre (with stores, bars and restaurants), the new Aquarium, a movie theatre complex and an IMAX cinema. The amount of building in this area is quite considerable; this has caused criticism with regards to the legitimacy of constructing such a singular structure right on the seafront, in addition to the fact that it has great accessibility problems.

d   *The new urban shopping centres.* Finally, the city has decidedly bet on a strategy to prevent commercial activities from moving to the surrounding parts of the metropolitan area. Thus, traditional commercial spaces have been re-equipped and integrated, giving them a new image and homogeneous promotion (Barnacentre, the commercial axis of Sant Andreu, Creu Coberta). And, on the other hand, just placing big shopping centres in the metropolitan rings, alongside major traffic arteries, has been avoided. Rather, there has been an attempt to anchor some of these centres inside the central city (Illa Diagonal, la Maquinista, Glóries, Heron, Diagonal Mar), by using vacant sites.

New productive activities and the 22@ Sector

Another intervention showing a clear will to defend the values of compactness and complexity in urban development is the 22@ program in the area of Poblenou. This is a district which was traditionally the city's eastern industrial centre (the other is found in the west, towards the Delta del Llobregat). In this area there has been a modification of the General Metropolitan Plan (passed on 27 July 2000) in order to allow for the establishment of new activities and the combination of uses. The old zoning of industrial land has been transformed into a new category, known as 22@, which, in an innovative manner, aims to attract activities linked to emerging sectors in manufacturing and creative industries (software production, telecommunications, multimedia, press, data-processing and electronic commerce, artistic activity and centres of investigation and education).

These new activities should generate, in the following years, between 100,000 and 130,000 new jobs, which could be combined with residential use of the same area. This is an attempt to attract emerging productive activities, alternating between industry and service, towards the centre of the city. This would establish them in an area in which, breaking the rigid schemes of zoning, they would be mixed with residential areas and centres of artistic production, education and research.

The magnitudes of the operation are as follows:[22]

- the area affected by the zoning of 22@ contains 198.3 hectares;
- the transformation of 1,159,626 m² of industrial land is foreseen, with a total potential of approximately 3,500,000 m² of new construction (excluding public facilities);
- the floorspace for new economic activity will be 3,200,000 m²;
- 4,614 homes that currently exist on industrial land will be recognized and an additional 3,500 new ones will be built with official (subsidized) protection;
- the enlargement of green zones will cover about 75,000 m² and the land destined for new facilities 145,000m²;
- this programme's potential in market value terms, at 2001 prices, is 7,813.2 million euros.

The future of the city and the Forum 2004

The above two interventions are mainly related to defending the values of compactness and complexity; the last one we wish to refer to has an added strong social aspect, intended to affect the city's social cohesion. We are referring to the Forum 2004 program for the coastal front of the Besòs.

The area constitutes a real compendium of urban problems, in which the following elements are all combined: large energy and environmental infrastructures (a water treatment plant, five power stations, a waste incineration plant), the mouth of a river, the Besòs, which is highly polluted, extensive road infrastructure (the recently finished Diagonal, the Cinturón del Litoral), mass housing estates from the 1960s and 1970s with significant social problems (la Mina, Sant Ramon de Penyafort, la Catalana), and the existence of the administrative boundary between Barcelona and the neighbouring municipality of Sant Adrià del Besòs.

The magnitude and the diversity of these problems have led to the design of a very complex programme, which includes the application of diverse specific projects and the organization of an important international event. Barcelona, deprived of the status of capital city and often ignored from the point of view of public investment, has turned many times to the resource of organizing a major event in order to give momentum to urban transformation during the past 150 years of its history. This is why it hosted the Universal Exhibitions of 1888 and 1929, the Eucharistic Congress of 1952 and the Olympic Games of 1992.

In this case, we are referring to an event conceived by the city of Barcelona itself, under the name of "Universal Forum of Cultures, Barcelona 2004". The Forum aims to be a

> new type of global event, organized in order to promote dialogue between peoples of all cultures. Debates will be organized on many topics of common interest, as well as a thematic World Festival of the Arts and exhibitions on human diversity and its history. It will be a major event, an expression of the creativity of all of the peoples and a political gathering of global dimensions with new imaginative forms of participation, centred on one of the main challenges of the twenty-first century: the dialogue between cultures.

This event, whose ambition cannot be denied, will take place during the summer of the year 2004 and has received the recognition and support of UNESCO.[23]

From the point of view of urban interventions, this plan affects a total of 170 hectares, practically adjacent to the 22@ district which was mentioned earlier, and includes the following operations:

- the construction, out into the sea, of a platform that will constitute the new seafront of the area, with beaches, a seaside promenade and an urban park, and to which part of the city's zoo will be moved from the Ciutadella Park;
- the construction of a new port next to the mouth of the river Besòs, with capacity for about 2000 boats;
- the construction of the installations of the Forum, with a central building, a large plaza and a platform that will cover the sewage treatment plant and, in part, the ring road of the Cinturón del Litoral;

- the restructuring of the neighbourhood of la Mina, and of the sectors of Llull-Taulat and la Catalana, with the support of the European Union's URBAN programme, and the creation of a special cross-authority committee;
- the reorganization of waste treatment and energy production infrastructures;
- the establishment of major metropolitan facilities, amongst which we find the Congress Centre of Catalonia and a new university campus, partially dedicated to biomedical scientific investigation.

The operation's success still showed unpredictable aspects, at the time of writing. It is, though, something new, with which Barcelona will try to combine economic innovation, urban renovation and social cohesion; that is, face some of the main challenges of the city's future.

Notes

1 For population and economic data on Barcelona, see *Barcelona Economia*, quarterly published by the Municipality of Barcelona (www.publicacions.bcn.es/bcneco), and the statistics provided by the Catalan Institute of Statistics (www.idescat.es).
2 Nello, O. (2002) *Cataluña, ciudad de ciudades*, Lleida: Milenio,159 pp.
3 Serra, J. (ed.) (2002) "Grans aglomeracions metropolitanes d'Europa", in *Papers. Regió Metropolitana de Barcelona*, 37, June, English text included.
4 Mancomunitat de Municipis de l'Àrea Metropolitana de Barcelona, *Dinàmiques territorials a l'àrea i regió de Barcelona*, Barcelona: Mancomunitat de Municipis de l'Àrea Metropolitana de Barcelona, 336 pp. (English text included).
5 Nello, O. (1998) "Spain", in Leo van den Berg *et al.*, *National Urban Policies in the European Union. Responses to the Urban Issues in the fifteen Member States*, Aldershot: Ashgate, pp. 325–77.
6 Clusa, J. (1996) "Barcelona: Economic Development 1970–1995", in Harris, N. and Fabricius, I. *Cities and Structural Adjustment*, London: University College of London, pp. 100–16).
7 Institut d'Estudis Regionals i Metropolitans de Barcelona (2002) *Enquesta de la regió de Barcelona 2000: primers resultats*, Barcelona: Institut d'Estudis Regionals i Metropolitans de Barcelona, 71 pp.
8 Clusa, J. and Cladera, J. R. (1999) "Evolució de l'estructura urbana de Catalunya, 1991–1996. Impacte dels canvis experimentats en la distribució de la població i la mobilitat per treball en el sistema català de ciutats", in *Nota d'Economia*, 64, pp. 66–94.
9 Institut d'Estudis Regionals i Metropolitans de Barcelona (2001) *Enquesta de la regió de Barcelona 2000: primers resultats*, Barcelona: Institut d'Estudis Regionals i Metropolitans de Barcelona, 71 pp.
10 Nello, O. (2002) *Cataluña, ciudad de ciudades*, Lleida: Milenio, 159 pp.
11 Serra, J. (ed.) (2002) "Grans aglomeracions metropolitanes d'Europa", in *Papers. Regió Metropolitana de Barcelona*, 37, June (English text included).
12 Institut d'Estudis Regionals i Metropolitans de Barcelona (2001) *Enquesta de la regió de Barcelona 2000: primers resultats*, Barcelona: Institut d'Estudis Regionals i Metropolitans de Barcelona, 71 pp.
13 Serratosa, A. (1994) "Els espais oberts en el planejament metropolità: realitats i propostes", in *Papers. Regió Metropolitana de Barcelona*, 20, pp. 37–47.

14 Nello, O., López, J. and Piqué, J. M. (1999) *Anàlisi de la mobilitat obligada dels municipis de la província de Barcelona, 1986–1996*, Bellaterra: Institut d'Estudis Metropolitans de Barcelona.
15 Institut d'Estudis Regionals i Metropolitans de Barcelona (2001) *Enquesta de la regió de Barcelona 2000: primers resultats*, Barcelona: Institut d'Estudis Regionals i Metropolitans de Barcelona, 71 pp.
16 Barcelona Regional (2001) *Criteris per al planejament metropolità de Barcelona*, Barcelona: Barcelona Regional, multicopied.
17 Institut d'Estudis Regionals i Metropolitans de Barcelona (2001) *Enquesta de la regió de Barcelona 2000: primers resultats*, Barcelona: Institut d'Estudis Regionals i Metropolitans de Barcelona, 71 pp.
18 Nello, O. (2003) "La ciutat de l'esperança", in *Saló d'Esperiències i projectes urbans*, Barcelona: Ajuntament de Barcelona.
19 *Barcelona economia*, various issues.
20 Institut d'Estudis Regionals i Metropolitans de Barcelona (2001) *Enquesta de la regió de Barcelona 2000: primers resultats*, Barcelona: Institut d'Estudis Regionals i Metropolitans de Barcelona, 71 pp.
21 Institut d'Estudis Regionals i Metropolitans de Barcelona (2001) *Enquesta de la regió de Barcelona 2000: primers resultats*, Barcelona: Institut d'Estudis Regionals i Metropolitans de Barcelona, 71 pp.
22 Ajuntament de Barcelona (2001) *Districte 22@*, Barcelona: Ajuntament de Barcelona. See, also, the website www.bcn.es/22@bcn
23 The official website of the Forum is www.barcelona2004.org

Chapter 3

Behind Barcelona's success story — citizen movements and planners' power

Nico Calavita and Amador Ferrer

Economic restructuring and deindustrialization have increased economic competition and pushed cities toward seeking more aggressive ways to effect economic development. The challenge facing cities — especially the older European and US industrial cities — is to respond in ways that will foster economic development while at the same time maintaining or improving the quality of life for their inhabitants in general and the poor in particular.

The city of Barcelona seems to have been successful in transforming itself in a very short time from a "grey" industrial city in the midst of a deep economic crisis in 1980 to an international success story a decade later. Most importantly here, Barcelona's city marketing effort was not accompanied by the neglect of its neighbourhoods, increased social polarization, or geographic segregation, distinguishing it in this regard from most other cities. How was all this accomplished?

The common perception is that skilled political entrepreneurship and creative planning unleashed through the Olympics enough money, energy and spirit of cooperation to do in ten years — with Barcelona preparing its bid as early as 1981 — what would normally have taken a much longer time.[1] In this chapter, we will show that the foundations on which the Olympics grew were laid down during the preceding decade. The 1970s were, without doubt, the most turbulent decade in Barcelona's post-Second World War history, comprising the last years of Franco's dictatorship of 36 years (known as "Franquismo") and the difficult years of transition to democracy.

It is important to note that Franco's dictatorship was not an anomaly in the history of nineteenth- and twentieth-century Spain, which was marked by an alternating succession of short democratic periods and longer military dictatorships. Franco's exceptionally long, personal dictatorship came after the six-year Spanish Republic and the subsequent Civil War of 1936–39. The result was the ruthless rubbing out of any form of public discussion and expression of individual opinions. During the last years of Franco's life, with the emergence of a new generation that had not participated in the Civil War, it became possible for a modicum of debate to take place, especially at the local level, with collective consumption and planning problems becoming the major topics of discussion and debate. We will argue that, during the 1970s, unique cultural, historical and political circumstances gave rise to collective urban social movements on one hand and exceptional individual progressive planners on the other, synergistically creating the conditions on which the Barcelona of the 1980s could be built.

The synergy and the conditions on which it was predicated can be divided into two periods, before and after the death of Franco in 1975. During the first quinquennium of the 1970s, the end of Franquismo was in the air and the urban movements, which already in the late 1960s had begun to organize to demand the public facilities sorely lacking in their neighbourhoods, acquired greater force and became part of the growing political opposition to the regime (Figure 3.1). In the planning arena, during these years a first version of the General Metropolitan Plan (GMP) of Barcelona was prepared and approved in 1974. This plan caused strong reactions on the part of citizen groups and especially property owners who saw the possibility for land speculation severely curtailed by the plan. While some of the changes requested by the property owners were conceded, the general features of the plan were maintained thanks to its defense by citizens groups.

The GMP of Barcelona (see colour plates 2 and 3), finally approved in 1976, marked the end of unbridled land speculation and the beginning of planning in the public interest. What is striking about this process is that it happened before a democratic administration took control of the city. Barcelona is, in fact, the only city in Spain where this change did not have to wait until the democratic administrations were in place.

The plan began to be implemented during the second half of the 1970s, as the slow and difficult period of transition to democracy unfolded, with the first local election taking place in 1979, after the general elections of 1977. During this uncertain and confusing time, the presence of a strong citizen movement made it possible to defend the plan against the old forces of speculation that wanted to build on lands the plan designated for public use. Equally important was the initiative displayed by the new planning director of the city of Barcelona, Joan Antoni Solans, appointed by the mayor of the "transition" period, who took advantage of the uncertainties in the urban land market to buy, at low prices, much of the land designated for public facilities in the GMP.

3.1
Front page of neigbourhood magazine 1974

With the advent of democracy, a new planning director, Oriol Bohigas, quickly responded to the demands of the neighbourhoods by seizing the opportunity offered by the newly acquired land and designed and built almost two hundred parks, plazas, schools and other public facilities. It was at this juncture, when many of the needs of the neighbourhoods were taken care of, that the city could turn its energy to solving its citywide needs — many contemplated in the GMP — such as opening the city to the sea, completing its inner and outer beltways, and expanding its drainage and sewer systems — to create the Olympic Barcelona.

Barcelona's urban development: a brief history

Barcelona is the capital city of the autonomous region of Catalunya in Spain. Founded as a Roman colony in the fifth century BC, Barcelona was able to quickly push out the Arab invaders from its territory in the eleventh century, becoming one of the most powerful maritime forces in the Mediterranean during the Middle Ages. Fiercely independent, Catalunya became part of the unified Spain in the late fifteenth century, setting the stage for innumerable struggles during the next five centuries with the central government in Madrid.

In the mid-nineteenth century, Madrid gave permission to tear down the walls that encircled Barcelona, no longer considered necessary for defence purposes — a process that was repeated in countless other European cities. The subsequent plan of expansion, the *Eixample*,[2] prepared by the engineer Ildefons Cerdà, represents one of the most significant examples of early modern town planning. His scheme was the first use of "neoliberal" plans that restructured or expanded existing cities during the second half of the nineteenth century to accommodate the commercial, social and representational needs of the new entrepreneurial class.[3] The *Eixample* became the framework within which the combined talents of Barcelona's turn-of-the century modernist architects could flourish. Their buildings, most notably Antoni Gaudí's, made Barcelona renowned as *La Ciudad de los Prodigios*, or the City of Marvels.[4] This was also a city of intense class conflicts, a city with no nobility and commoners but of capitalists and workers, many of them anarchists and Marxists. The city was often rocked by episodes of revolt aimed at the church and the clergy.[5]

After the turn of the century, the idea of an *Exposicion Universal* that would act as a showcase of Barcelona's status as the prominent manufacturing centre in Spain took hold. One of the most important architects of the modernist period, Josep Puig i Cadafalch, also a politician, supported the idea as a mechanism to improve the infrastructure of the city. While the main buildings of the Exposicion of 1929 were located on the mountain of Montjuic, the entire city benefited from a system of avenues, parks, buildings and plazas built at this time.

With the advent of the 1931 republic, Cataluna was granted autonomy, inaugurating an intense, albeit short-lived, period of intellectual and artistic activity. The architects of this period became aligned with the ideas of the modern movement in architecture and urban planning that had swept Europe during the previous two decades. They formed the Grup d'Artistes y Técnics Catalans per al Progres de l'Arquitectura Contemporánia (GATCPAC). The leader of the group was Josep Lluis Sert, who was to become a major figure in the history of modem architecture for his work in both Spain and the United States. GATCPAC prepared a plan for the Barcelona metropolitan area in 1934, the so-called Pla Macià after the name of the president of the region. Le Corbusier, arguably the most prominent architect in the history of modern architecture and urban planning, collaborated with the young Catalan architects. Not surprisingly, the plan reflected Le Corbusier's ideas of functional separation of uses and the use of superblocks. But the plan remained on paper as the Civil War put an end to the creative impulses of this period. With the victory of Franco in 1939, Cataluna lost its autonomy and its connection to the European avant-garde, entering a 36-year period of repression and obscurantism.

In the economic sphere, Franco pursued a policy of isolationism and autarchic policy until the year 1959, a year when the economy was liberalized, leading to an unprecedented period of economic growth during the 1960s and early 1970s.[6] As a result, many Spanish cities, especially those located in the industrial regions of the country, went through an unprecedented process of growth. At the end of that process, those cities had become large metropolitan areas, heavy-weight players not only in the Spanish but also in the European context.

The areas of new growth followed three different patterns of development that even today define the structure of the residential peripheries of most Spanish cities. These peripheries constituted the cultural medium through which urban social movements developed.[7] We can distinguish three patterns of development.

1   Suburban developments based on nineteenth-century extensions with narrow streets that were massively densified, substituting, for example, the original two-floor small house with a rear garden located on a 6 m wide lot with six- or seven-floor blocks of apartments. These changes led quickly to the appearance of functional and formal conflicts: lack of facilities due to the sharp increase in population, car access difficulties, and poor lighting and ventilation conditions.
2   The so-called marginal areas of urbanization, located in the extreme periphery of the city and illegally built. This pattern of urban growth had its origin in the 1920s — a period of high growth especially in Barcelona — and became significant in the 1950s in cities like Barcelona, Madrid and Bilbao. These shanty towns lacked all public facilities and became a strong focus of urban conflicts.

3   The new housing projects, or poligonos de viviendas, most of them publicly promoted to house low-income families, appeared first in the 1950s and then mushroomed during the 1960s and early 1970s as the official answer to the housing shortage. As a consequence of the priority to house more and more people, the provision of public facilities for these projects was systematically neglected. Furthermore, the poligonos were often located in isolated settlements and poorly built with densities usually higher than 70 apartments per acre and as high as 150 in some cases.[8] It is not surprising that they generated the sharpest conflicts.

This continuous process of densification added a large residential mass to Barcelona and the surrounding municipalities, transforming the area into an integrated metropolis that grew by 1,500,000 inhabitants between 1950 and 1980.[9] At the same time, this process of growth fostered "an incomplete urban periphery because of a lack of services, centrality and/or symbolism and image."[10] The resulting progressive increase in social segregation and urban conflicts led to the urban social movements.

Urban social movements

Manuel Castells, in his classic *The Urban Question* published in 1972, challenged conventional urban sociology that negated social conflicts and proposed instead a new definition of the urban problem based on the theory of "collective consumption" — goods and services directly or indirectly provided by the State — and/or urban social movements that, in the struggle to get their fair share of those goods and services, become the catalyst of the transformation of the social relations of society. Urban social movements, then, are not necessarily concerned with class or workplace issues but primarily with neighbourhood concerns, that is, with the organization of the urban space shared by the members of a movement and involving conflict with the State over the provision of, access to, or defence of, urban services and housing.[11] They were, in other words, "the symptom of a process of social readjustment," and the change necessitating readjustment was rapid urbanization.[12]

In Spain these movements, it should be remembered, took place in a particular political context, the final period of Franquismo until 1975, and of the transition to democracy during the remainder of the decade. Having to operate under those conditions, the Spanish movements faced many difficulties, but at the same time, the dictatorship served as a concrete and highly visible target for the opposition. Also, it should be mentioned that the repressive nature of the Franco regime weakened during the last years of the regime, possibly due to the political mobilization of workers, regionalists and students, in addition to the urban social movements.[13]

In the Spanish context, urban social movements were characterized by:

1   direct action and protest tactics focused on issues of collective consumption;
2   a grassroots orientation; and
3   a certain distance from political parties (clandestine until the mid-1970s).[14]

The Barcelona case

In Barcelona, the neighbourhood associations (*Asociaciones de Vecinos*) came about in response to everyday problems specific to particular neighbourhoods, "to win a set of traffic lights, to have some running water in the houses, to have drains, asphalt, to put an end to the dust and dirt in the streets". These problems compounded "the already hard working conditions, low salaries and long hours spent commuting to work."[15] At the end of the 1960s, the urban social movements grew quickly in parallel with the rapid urbanization and densification of the city, making visible the shortages of public facilities and the neglect of urban space in the new and old urban peripheries.

The initial forms of protest included the collection of signatures, assemblies, expositions, gatherings around sport or music events, symbolic inaugurations, and so forth. The habitual response of the administration was silence, leading to harder forms of struggle in a context where social liberties were non-existent. The lack of social liberties made organization difficult, but it meant much more:

> getting the authorities to give us a set of traffic lights meant forty days of barricades and stopping cars coming into the district where four or five fatal accidents had taken place. And that meant clashes with the police. Demanding mains for the shacks in Torre Baro meant cutting off the motorways into Barcelona every day, with everything that involved. It was a difficult time.[16]

Other forms of struggle included occupation of public spaces, human barriers, sequestering buses and rent strikes.[17]

Protest actions in individual neighbourhoods were not isolated actions; they were part of a wider protest movement at the city level. While the urban social movement generally acted separately from political parties, in Barcelona at the end of the 1960s and the beginning of the 1970s there existed an intangible network between the protests of the neighbourhoods, the activities of the clandestine unions and the forbidden political parties, the university protests and the activities of professional organizations.[18]

Individual intellectuals and (still clandestine) politicians became involved as well. Surreptitiously backed by a significant portion of the press, the Barcelona social movement quickly became an alternative forum for the discussion of urban affairs.

By the early 1970s, the scope of the neighbourhood associations expanded to include urban planning issues such as the need for public facilities, especially areas for open space and parks, and opposition to "Partial Plans." These *Planes Parciales* were a tool to supplement the existing 1953 Comarcal Plan — a plan that was too general and required Partial Plans and accompanying ordinances to be implemented.[19] These plans increased densities without a corresponding provision of public facilities — at times preempting spaces dedicated to public facilities — and had been used to satisfy the interest of speculators during the long administration of Mayor Josep Maria de Porcioles. His 16-year mandate (1957–73) coincided with the huge demographic change mentioned above, and the partial plans became "the root of all speculation" by always allowing "increases in permitted building levels, granted by an absolutely permissive government, even approving housing estates without preliminary partial planning."[20] The result was termitaries built by speculators under license from the Caudillo's placemen, designed without paved roads, playgrounds for the children, or other signs or thought for infrastructure or public space, quite often made of poor materials that started falling apart within a few years.[21]

The Barcelona of those years has been dubbed *Barcelona Grisa*, "grey Barcelona", and *porciolismo*, the abandonment of the city on the part of administrators in the hands of speculators.[22]

In 1973, 20 years after the adoption of the Comarcal Plan, Porcioles's mandate ended, the result of an attempt to pass a partial plan that would have destroyed 4,370 homes. The neighbourhoods involved came together to fight this threat, and the ensuing struggle culminated in a confrontation in city hall at the time the partial plan was to be approved. Before the meeting began, municipal functionaries occupied the seats usually reserved for the public to preclude the participation of protesting citizens. In response, the neighbourhood associations occupied city hall, the facts were publicized, and the resulting scandal forced the central government in Madrid to depose Porcioles on the following day. The plan was stopped, and the neighbourhoods involved came together to form an association called *Nou Barris*, "nine neighbourhoods," which many years later became one of the ten districts into which Barcelona was divided as part of the administrative decentralization of the city.

This episode is important as a sign of the weakening grip of Franquismo on the country, of the power of the Barcelona urban social movement, and of the expansion of its reach to include large-scale planning issues. It is within this atmosphere that the preparation of the GMP began (Figure 3.2).

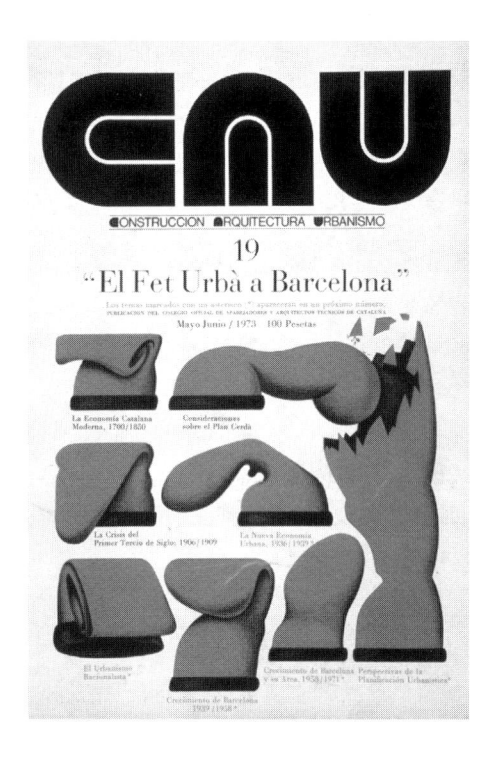

3.2
Front page of architects' publication, 1973

The General Metropolitan Plan

By the late 1960s, it had become clear that the 1953 *Plan Comarcal* needed to be updated. The new *Comarcal Plan*, initially approved in 1974 and finalized in 1976, was an exceptionally advanced plan.

That such a plan could be produced in a Spanish city still under the Franco regime is nothing short of miraculous. The political, social and cultural rupture with the rest of Europe resulting from the Civil War had led to a cultural isolationism that truncated some of the planning experimentation that had aligned Barcelona architecture and planning with the modern movement.[23]

In the 1960s, however, a new generation of architects[24] opened "itself to the world and joined the rest of Europe".[25] This cultural awakening was made possible by the weakening hold of the regime on the cultural life of the nation and the normalization of book and journal imports from abroad. Books by Lynch (1960), Jacobs (1961), Chermayeff (1962), Eco (1962), Alexander (1963), Venturi (1965), Rossi (1966) and others led to a rejection of the doctrinalism into which the modern movement had fallen. US publications introduced Barcelona architects to the criticism of the Urban Renewal program and to Advocacy Planning. Also, the ferment in European architecture schools, especially in France and Italy, led to an analysis and criticism of the political-economic context of planning and clarified "the social and political compromises that underlie much of the planning practice of an important portion of professionals".[26] However, it was Italian planning of the 1960s that had the greatest influence on Barcelona architects, especially the "Piano Intercomunale Milanese," the plan for the Milan metropolitan area, presented in the October 1967 issue of *Urbanistica* that provided, in its break with tradition, the most important disciplinary lesson for the team in charge of the Plan Comarcal.

Following the Piano Intercomunale Milanese, the Plan Comarcal rejected the traditional, architecturally-based approach to planning to emphasize "*una politica del suelo*" (meaning land policy, or perhaps more accurately the politics of land):

> for the first time, there is a coherent plan that establishes intensities and densities of development ... based on the introduction of legal controls that regulate the growth of the city.[27]

The innovation of the plan at the political level can be narrowed to two elements:

1   it reduced the allowable densities from a potential of nine million people to four and one-half; and
2   it reclaimed land for public use by designating various parcels of land for parks, plazas, schools and other public facilities.[28]

About half of the land designated for public use under the previous plan had been used for speculative housing projects.

The plan, prepared by a team headed by an engineer, Albert Serratosa, an architect, Joan Antoni Solans, and an attorney, Miquel Roca Junyent, was first unveiled in 1974. During the same year, the *Corporacio Metropolitana de Barcelona* was created as the entity that included Barcelona and 26 municipalities. It covered the same territory as the *Comarca*, but its new metropolitan reality was recognized and the plan became the GMP. A new mayor, Enric Masó, appointed to fill the post left open by the unceremonious dumping of Porcioles, had been in office only a few months when the plan was unveiled. Direct appointment of mayors, formally by the Spanish central government (but really by the dictator, among a short list of suggested candidates presented by the government), was the normal practice for the big Spanish cities during the Franco regime. In smaller cities nominations were controlled by the provincial governors. The mayor of Barcelona was also in charge of the *Corporació*.

The new mayor attempted to establish a dialogue with the neighbourhood associations and made regular visits to the most degraded neighbourhoods. However, when the new GMP was unveiled a few months after his inauguration, it was attacked by the *Asociaciones*, both because they felt that not enough areas had been designated for public use and because the proposed new thoroughfares — "a personal fixation" of the

3.3
Via Júlia in 1970s

engineer Serratosa[29] — would have cut through some of the historic neighbourhoods, such as Gracia, and affected thousands of homes. There were many errors as well, with parcels designated for public use that the *Asociaciones* quickly discovered were being or had been already built upon. The 1974 plan became the vehicle through which the various citizen movements were agglutinated and consolidated, strengthening their resolve to obstruct further deterioration of their city (Figure 3.3).

But an even more powerful attack came from the landowners who saw the potential volumes to be built on their properties drastically reduced and/or their land designated for public use, sharply devaluing their properties. Even though the changes had become necessary to ameliorate the damages inflicted on the city by 20 years of *porciolismo*, the landowners closed ranks against the plan. This explains why the new plan was denounced by one of the main developers in Barcelona as "a socialist plan,"[30] or that a banker lamented in *La Vanguardia Espanola* that the loss of land value under the new plan could reach 500,000 pesetas. He also noted that:

> the planning process which has led to this plan is extremely curious, with no public involvement …, has been a thing of very few people …, an engineer with political aspirations; a recently graduated architect who has studies abroad, a lawyer, and very little more.

A little later he dubbed them "young messiahs" and a "fistful of technocrats."[31]

The plan gave rise to such passionate conflicts that the central government decided "to substitute a conciliatory mayor with an intransigent one."[32] The new mayor, Joaquim Viola, was a friend of Porcioles. As the plan was being revised, hundreds of building permits were released under the aegis of the 1953 plan, leading to "the unfortunate loss of parcels important for the future implementation of the GMP."[33]

Landowners pressed for change to the new plan as well, and the new mayor, closely connected to the old power structure, was sympathetic to changes favourable to his cronies. Serratosa, "the father of the Plan," was subjected to innumerable pressures and threats and had his children accompanied by bodyguards to school.

The neighbourhood groups rose to the occasion and took to the streets, demanding the elimination of roads that cut through their neighbourhoods, and the redesignation to public use of areas that had been changed to private use, such as the *Espana Industrial*, a huge complex of textile factories that had ceased to operate. They were supported by professional associations, such as that of the architects. The thoroughfares were eliminated. But many of the other objections of the *Asociaciones* were not met when the GMP was approved by Viola in the summer of 1976. The *Asociaciones* realized that their chances of success would be nil with Viola in power and turned their energies toward eliminating him from the scene. They demanded his resignation from the king, the minister of internal administration, and the governor of Barcelona. Posters appeared on the neighbourhood walls demanding his resignation.

In December 1976, Viola resigned, and a new "conciliatory" mayor, Josep Maria Socias, took his place. The battle for the plan continued through the courts, where the *Consejo de Estado* fixed some of the worst excesses of the approved GMP. The *Espana Industrial* site, for example, was redesignated parkland, raising the ire of its owner, who wished a coup d'état à la Chile upon Spain.[34] A few years later, Serratosa, commenting on the role that the neighbourhood associations played in the revision of the GMP declared that they "were the real protagonists ... in resisting the attacks on the most essential aspects of the plan on the part of powerful pressure groups."[35] Much later he credited the citizens' defence of the GMP for "permitting a bridge between technique and politics, which is one of the important impediments to planning, even today."[36]

The new mayor Socias took the reins of the city in the difficult years of transition to democracy. The first general election had taken place in 1977, but local elections had to wait until 1979. Until that time, Socias decided that to legitimize his government, he had to be more responsive to the neighbourhood associations. His mandate allowed a relaxation of the tensions created by his predecessor, Viola, and a smooth transition to democracy. It was a time full of life, energy, and dynamism, when the most difficult problems of the city were discussed with the will to find solutions. One of the first tasks of the city was to find solutions to the most difficult demands for public facilities like schools and parks.

Planning implementation, municipal election, and the demise of the *Movimiento de Vecinos*

The mayor decided to hire the other "father" of the GMP, Joan Antoni Solans, as planning director. Together, they decided to acquire as much as possible of the areas designated for public use. On one hand, this attempt was extremely difficult, given the conditions of the public administration after decades of Franquismo, and "the moral and physical conditions" of Barcelona at the end of the Viola administration. On the other hand, the moment was propitious because the political context of the transition period had created a situation of diffuse fear among landowners that the country and the city were inexorably moving toward socialism and the requisition or expropriation of their property at low prices. The economic recession of that period and a scarcity of capital for developers helped even more to bring the price of land down. The availability of funds was greater than initially projected because a large portion of the necessary funds was made available from the central government. In the end, Solans was able to buy almost 221 hectares (approximately 500 acres) of land at a cost of 3 billion pesetas: 86 hectares for parks and gardens; 50 for forest land, 70 for school sites and other public facilities; and 15 for housing.[37]

Despite the great insistence of the *Vecinos* for municipal elections, they did not take place until four years after Franco's death. The Left won, and a socialist, Narcis Serra, was elected mayor of Barcelona. Many of the members of the *Asociaciones de Vecinos* and of the professional organizations that supported them either were elected to city council or entered the new administration. The two main tasks of the newly elected government were to reform its public administration and respond to the strong social pressure for the planning *recuperacion* of the city that had been expressed during the last years of the old regime and during the transition period and that had found expression in the 1976 GMP plan that was adopted by the new administration.

With the advent of democracy, a new planning director, Oriol Bohigas, immediately began the process of planning and design for the newly acquired land. He was the catalyst who brought together a large number of young architects who had entered the profession during the 1970s to design almost two hundred parks, plazas, schools and other public facilities. Because of the recent approval of the GMP, Bohigas did not have to spend time to prepare a new plan for the city but could dedicate his energies to responding to the urgent need for new public facilities, especially new public spaces and services.[38] "Instead of working from the general to the particular, from master plan to local project, Bohigas inverted this usual procedure …, he focused instead, as and where opportunities presented themselves." To respond quickly to citizens' demands and to project an image of an efficient administration, Bohigas began to design and build first what could be done the fastest and the cheapest, seizing "not just local attention but also the international acclaim that helped fuel local enthusiasm … this shrewdly opportunistic strategy was then progressively escalated."[39]

It is at this time that the urban social movement in Barcelona — and other Spanish cities as well — lost much of its momentum, power and membership. There are several reasons for this sudden change. First, the urban social movement lost its most important *raison d'être* with the completion of many of the needed projects. Second, it should be remembered that the demands of the *Vecinos* had been part of a larger political opposition to the Franco regime. Now, with the democratization of the political system in general and a socialist administration in particular — composed to a large extent of former members or sympathizers of the *Asociaciones de Vecinos* — opposition withered. It is not surprising then that throughout Spain urban political movements came practically to a standstill, and the *Asociaciones* were decimated.[40] One of their hopes, that the new administration would install a more participatory form of democracy, remains unfulfilled. The president of the Federacion de Asociaciones de Vecinos Barcelona (FAVB), Carles Prieto, told a reporter in 1982 that "the political parties of the governing coalition have abandoned the *Asociaciones de Vecinos*."[41] Since then, the FAVB has been critical of all the Left administrations that have continued to govern the city. Ironically, the FAVB continues to operate in offices and with funds provided by those same administrations.

The fate of the Spanish urban social movements and studies of other urban social movements have confirmed their contingent nature.[42] Hindsight has made it clear that Castells overestimated the long-term impact or endurance of the urban social movements. Certainly, social movements have not, as Castells theorized, fulfilled their supposed role as agents of social change or heralds of a new era. "It is possible," as Ceccarelli explains, "that urban social movements and the conflicts which accompanied them, far from predicting a new era, were the expression of the last and most conflictive stage of a process of change and readjustment to it."[43]

Conclusion

The first thing that the new mayor, Narcís Serra, did when he took office was to respond to the needs of the neighbourhoods. But as that process got under way, he expanded his sights to problems at the citywide level. When presented with the idea of the Olympics, he seized the opportunity. Thus began the process that led to the opening of the city to the sea, building a system of inner and outer beltways, and repairing and expanding its drainage and sewer systems, to name the most important projects (see colour plate 4). In 1984, the noted architectural critic Peter Buchanan wrote an article in the *Architectural Review*, entitled "Regenerate Barcelona," on nine of the most impressive parks and plazas being built at that time in Barcelona. In 1992, another article in the same journal, entitled "Barcelona Regenerated," described the changes brought about by the 1992 Olympics. Barcelona had been placed on the map, boasted the next mayor Maragall.

It would be tempting to ascribe, at least in part, the success of Barcelona's Olympics to a citywide and neighbourhood participation process in its planning and implementation. But this is not the case. In fact, the FAVB lamentations about the lack of citizens' involvement in the affairs of the city are particularly sharp in their criticism of how the handling of the Olympics was a private affair between the "prince," and — as in the Renaissance — the "architects of the prince."[44] This criticism of the FAVB might be too harsh. The "Prince" in question is Pasqual Maragall, the mayor who inherited city hall from the first democratically elected mayor, Narcís Serra (who moved to the Ministry of Defense in Madrid) and with it, the organization of the 1992 Olympics. He had a mere ten years to obtain the financing, to plan, and to build not only the Olympic facilities but the infrastructure of the city that had been left in shambles after 20 years of *Porciolismo*. Relying on a handful of technocrats — mostly architect/planners and engineers who had emerged during the turbulent 1970s — he achieved what many doubted could be accomplished in such a short time. It is debatable whether a more participatory process would have been possible without jeopardizing the timely completion of all the facilities necessary for the Olympics.

The dream of the FAVB and others for a more participatory form of government has not materialized.[45] Nevertheless, what is remarkable is that the city's attempt at city marketing — of which the Olympics were an essential ingredient — has not hampered, as is generally the case,[46] the pursuit of an ambitious social agenda. In other Olympic cities, such as Atlanta, for example, the promise of neighbourhood regeneration has remained generally unfulfilled.[47] In Barcelona, the city took care of the neighbourhoods first, as part of the overall strategy, established soon after the 1979 election, that sought the "homogenization of the city," that is, "the creation of a balanced and integrated Barcelona, without segregation, with social and territorial equality for all its citizens."[48] Such a strategy, established immediately after the 1979 election, did not materialize out of thin air but was the result of the legacy of the 1970s.

Not only did it bring balance in the provision of public facilities throughout the city, thus helping to integrate the marginalized neighbourhoods, but it has also reversed the incipient tendencies toward segregation in the historic centre and some of the peripheral areas.[49] Income inequality seems to be lessening as well. A recent survey has found that between 1985 and 1995, income inequality has decreased in the city.[50]

In June 1999, the Royal Institute of British Architects (RIBA) awarded Barcelona its prestigious Gold Medal — the first time awarded to a place and not professionals — in honour of the city's "commitment to urbanism over the last 20 years" including its "mix of eye-catching landmark projects, small-scale improvements to plazas and street corners, and the team work between politicians and urbanists."[51] While the fruits have matured during the past 20 years, we should not forget that the seeds were sown earlier, that Barcelona's success in balancing economic development with the enhancement of its quality of life is based on what happened in the 1970s — on the courage and commitment of the *Vecinos* and the planners who laboured under harrowing conditions for a better quality of life for all the people of Barcelona.

Notes

1   Buchanan, P. (1992) "Barcelona, A City Regenerated", *The Architectural Review*, August. Moix, L. (1994) *La Ciudad de Los Arquitectos*, Barcelona: Cronicas, Anagrams.
2   *Exiample* is the Catalan translation of *expansion*. Catalan is the official language of Catalonia. Franco, after his victory in 1936, made Catalan illegal and imposed Castillian (Spanish) on the Barcelonese. In this chapter, we will use the language as it appears in the literature used or as it was used in a particular period. Ensanche is the Spanish word for expansion.
3   Benevelo, L, (1967) *The Origin of Modern Town Planning*, Cambridge: MIT Press. The most well known is Baron Haussman's plan for Paris, implemented during the reign of Napoleon III.
4   Hughes, R. (1993) *Barcelona*, New York: Vintage.
5   Hughes, R. (1993) *Barcelona*, New York: Vintage.
6   Ferrer, A. and Nello, O. (1990) "Barcelona: The Transformation of an Industrial City," in Salamon Lester (ed.), *The Future of the Industrial City*, Baltimore: Johns Hopkins University Press.
7   Busquets, J. (1992) *Barcelona*, Madrid: Editorial Mapfre. Sola-Morales, M. *et al.*, (1974) *Las Formas de Crecimiento Urbano*, Barcelona: LUB ETSAB.
8   Ferrer, A. (1996) *Els Poligons de Barcelona*, Barcelona: UPC.
9   Ferrer, A. and Nello, O. (1990) "Barcelona: The Transformation of an Industrial City," in Salamon Lester (ed.), *The Future of the Industrial City*, Baltimore: Johns Hopkins University Press.
10 Busquets, J. (1992) *Barcelona*, Madrid: Editorial Mapfre. Sola-Morales, M. *et al.*, (1974) *Las Formas de Crecimiento Urbano*, Barcelona: LUB ETSAB , p. 257.
11 Pickvance, C. (1985) "The Rise and Fall of Urban Movements," *Environment and Planning D: Society and Space*, 3: 31–53.
12 Lowe, S. (1986) *Urban Social Movements. The City after Castells*, London: Macmillan. Pickvance, C. (1985) "The Rise and Fall of Urban Movements," *Environment and Planning D: Society and Space*, 3. This contingent nature of urban social movements is reflected in the nature of social movements in the United States in the same period, seen as a reaction, not to rapid urbanization, but to urban restructuring, that is, the combination of urban renewal, gentrification, and highway projects that were aimed at creating the post-industrial city. See Fainstein, S. *et al.*, (1983) *Restructuring the City*, New York: Longman. US neighbourhood movements were eventually successful in stopping urban renewal and freeway construction in their communities. See Mollenkopf, J. (1983) *The Contested City*, Princeton, NJ: Princeton University Press.
13 Castells, M. (1983) *The City and the Grassroots*, Berkeley: University of California Press.
14 Borja, J. (1977) "Urban Movements in Spain," in M. Harloe (ed.), *Captive Cities*, New York: John Wiley. Castells, M. (1983) *The City and the Grassroots*, Berkeley: University of California Press.
15 Naya, M. "The Neighbourhood Assocations," in Mateo, J. L. (ed.) (1996) *Comtemporary Barcelona 1856–1999*, Barcelona: Centre de Cultura Contemporania, p. 191.

16. Naya, M. "The Neighbourhood Assocations," in Mateo, J. L. (ed.) (1996) *Comtemporary Barcelona 1856–1999*, Barcelona: Centre de Cultura Contemporania.
17. Huertas, J. and Andreu, M. (1996) *Barcelona en Lluita: El Moviment Urbá 1965–1996*, Barcelona: Federació d'Associacions de Veins de Barcelona.
18. Vázquez Montalban, M. (1996) "La lluita necessaria," in *Barcelona en Lluita: El Moviment Urbá 1965–1996*.
19. The Comarcal Plan was prepared by a small group of planners with the city of Barcelona for the entire Comarca, an administrative unit that comprised 27 municipalities, with Barcelona being the dominant city. While technically adequate, the plan could not resist the population pressure that followed its adoption and became quickly obsolete, merely a reference for the partial plans that ignored its guidelines.
20. Salvadó, T. and Miró, J. "The Appendages of the City of Kidneys," in Mateo, J. L. (ed.) (1996) *Contemporary Barcelona 1856–1999*, Barcelona: Centre de Cultura Contemporania, p. 143. This ability of landowners to modify plans in their interests confirms Borja's (1997) representation of the local state as weak and open to external domination of the proprietied classes. This weakness is reflected also in the local government being "the least repressive sector of the state," Borja, J. (1977) "Urban Movements in Spain," in Harloe, M. (ed.), *Captive Cities*, New York: John Wiley, p. 190.
21. Hughes, R. (1993) *Barcelona*, New York: Vintage. Robert Hughes is the flamboyant former art critic of the *New York Times*. We assume that he uses "termitaries" or termites' nest, to portray dramatically the high densities of those buildings.
22. Such a term is still used in contemporary Barcelona. For example, a recent interview of a Barcelona architect critical of the present administration had the title: "A Sort of Porciolismo is Returning," in Gali, B. "Vuelve una especie de porciolismo' *El Pais*, 24 July 1998. Cataluna section, 4.
23. Busquets, J. (1992) *Barcelona*, Madrid: Editorial Mapfre.
24. In Europe, and particularly in Italy and Spain, architectural schools have generally also taught urban planning, and practically all architects consider themselves planners as well. In this chapter, architects will be considered, as in Spain, architect-planners. For a commentary on a similar situation in Italy, see Calavita, N. (1984) "Viewpoint", *Planning*, 50, No. 11.
25. Pié, R. (1996) "El Projecte Disciplinar: La Versio de 1974 Pla General Metropolita" in *Els 20 Anys del Pla General Metropolitá de Barcelona*.
26. Pié, R. (1996) "El Projecte Disciplinar: La Versio de 1974 Pla General Metropolita" in *Els 20 Anys del Pla General Metropolitá de Barcelona*.
27. Solans, J. A. (1996) "El Pla Metropolitá," in Mateo, J. L. (ed.), *Contemporary Barcelona 1856–1999*, Barcelona: Centre de Cultura Comtemporania, p. 202.
28. Land for public facilities at the subdivision level was to be obtained through the subdivision process; at the neighbourhood level, land for facilities was to be expropriated. Spanish legislation allows expropriation of land at below market value.
29. Huertas, J. "El moviment ciutada a Barcelona i l'aparició del Pla General Metropolitá," in Ferrer, A. (ed.) (1996) *Els 20 Anys del Pla General Metropolita de Barcelona*, Barcelona: Institut d'Estudis Metropolitans de Barcelona.
30. Samaranch, J. A. quoted in Huertas, J. "El moviment ciutada a Barcelona i l'aparició del Pla General Metropolitá," in Ferrer, A. (ed.) (1996) *Els 20 Anys del Pla General Metropolita de Barcelona*, Barcelona: Institut d'Estudis Metropolitans de Barcelona.
31. Trias Fargas, R., quoted in Huertas, J. "El moviment ciutada a Barcelona i l'aparició del Pla General Metropolitá," in Ferrer, A. (ed.) (1996) *Els 20 Anys del Pla General Metropolita de Barcelona*, Barcelona: Institut d'Estudis Metropolitans de Barcelona, p. 65.
32. Huertas, J. "El moviment ciutada a Barcelona i l'aparició del Pla General Metropolitá," in Ferrer, A. (ed.) (1996) *Els 20 Anys del Pla General Metropolita de Barcelona*, Barcelona: Institut d'Estudis Metropolitans de Barcelona.; Huertas, J. and Andreu, M. (1996) *Barcelona en Lluita: El Moviment Urbá 1965–1996*, Barcelona: Federació d'Associacions de Veins de Barcelona.
33. Busquets, J. (1992) *Barcelona*, Madrid: Editorial Mapfre.

34 Huertas, J. "El moviment ciutada a Barcelona i l'aparició del Pla General Metropolitá," in Ferrer, A. (ed.) (1996) *Els 20 Anys del Pla General Metropolita de Barcelona*, Barcelona: Institut d'Estudis Metropolitans de Barcelona.; Huertas, J. and Andreu, M. (1996) *Barcelona en Lluita: El Moviment Urbá 1965–1996*, Barcelona: Federació d'Associacions de Veins de Barcelona.

35 Quoted in Huertas, J. "El moviment ciutada a Barcelona i l'aparició del Pla General Metropolitá," in Ferrer, A. (ed.) (1996) *Els 20 Anys del Pla General Metropolita de Barcelona*, Barcelona: Institut d'Estudis Metropolitans de Barcelona, p. 68.

36 Serratosa, A. "La Revisiódel Pla Comarcal de 1953," in Ferrer, A. (ed.) (1996) *Els 20 Anysdel Pla General Metropoliá de Barcelona*, Barcelona: Institut d'Estudius Metropolitans de Barcelona, p. 14.

37 Solans, J. A. (1979) "Barcelona 1977–1978," *ARQUITECTURAS BIS*, No. 3: 51–2.

38 Buchanan, P. (1994) "Regenerating Barcelona: Projects versus Planning — Nine Parks and Plazas," *The Architectural Review*, June. Moix, La Ciudad de Los Arquitectos.

39 Buchanan, P. (1994) "Regenerating Barcelona: Projects versus Planning — Nine Parks and Plazas," *The Architectural Review*, June. Moix, *La Ciudad de Los Arquitectos*.

40 Castells, M. (1983) *The City and the Grassroots*, Berkeley: University of California Press; Pickvance, C. (1985) "The Rise and Fall of Urban Movements," *Environment and Planning D: Society and Space*, 3.

41 Quoted in Huertas, J. "El moviment ciutada a Barcelona i l'aparició del Pla General Metropolitá," in Ferrer, A. (ed.) (1996) *Els 20 Anys del Pla General Metropolita de Barcelona*, Barcelona: Institut d'Estudis Metropolitans de Barcelona; and Huertas, J. and Andreu, M. (1996) *Barcelona en Lluita: El Moviment Urbá 1965–1996*, Barcelona: Federació d'Associacions de Veins de Barcelona.

42 Dunleavy, P. (1980) *Urban Political Analysis*, London: Macmillan. Pickvance, C. (1985) "The Rise and Fall of Urban Movements," *Environment and Planning D: Society and Space*, 3; Lowe, S. (1986) *Urban Social Movements. The City after Castells*, London: Macmillan; Fincher, R. (1987) "Defining and Explaining Urban Social Movements," *Urban Geography*, 10: 604–13; McKeown, K. (1987) *Marxist Political Economy and Marxist Urban Sociology*, London: Macmillan; Purcell, M. (1997) "Ruling Los Angeles: Neighbourhood Movements, Urban Regimes, and the Production of Space in Southern California", *Urban Geography*, 18, No. 8: 684–704.

43 Ceccarelli, P. (1982) "Politics, Parties and Urban Movements: Western Europe," in Fainstein N. and Fainstein S. (eds), *Urban Policy under Capitalism*, Beverly Hills, CA: Sage, p. 264.

44 Federacion de d'Asociaciones de Vecinos Barcelona (FAVB), *La Barcelona de Maragall* (Edicions Sibilla).

45 Moreno E. and Vásquez Montalban, M., (1991) *Barcelona, cap a on vas?* Index Barcelona.

46 Sassen, S. (1991) *The Global City*, Princeton, NJ: Princeton University Press, 1991. Eade, J. (ed.) (1996) *Living the Global City: Globalization as Local Process*, London: Routledge.

47 French, S. and Disher, M. (1997) "Atlanta and the Olympics: A One-Year Retrospective," *Journal of the American Planning Association*, 3, No. 3: 379–92.

48 Casas, X. (1997) "La Transformacio del litoral del pla de Barcelona com a motor de la renovacio de la ciutat," *Ajuntament de Barcelona*, April 19, 5.

49 Casas, X. (1997) "La Transformacio del litoral del pla de Barcelona com a motor de la renovacio de la ciutat," *Ajuntament de Barcelona*, April 19. New housing projects publicly promoted by PROCIVESA (Promocio Ciutat Vella, S.A.) have broken up the social–age–income uniformity existing in the historic centre, by making it possible for middle-class, upper-class, and younger families to live in an area that was dominated by a lower-class, older population. New housing projects near the sea have also pursued income integration, and in the urban periphery, programmes are in place to avoid ghetto formation.

50 Nello, O. et al., (1998) *La transformacio de la ciutat metropolitana*, Barcelona: Area Metropolitana de Barcelona, Diputacio de Barcelona, p. 120. We do not want to give the impression of Barcelona as the "City on a Hill." Obviously, problems in Barcelona persist. High housing costs are pushing the younger generations to seek housing in the more

affordable towns outside of Barcelona. The FAVB is critical of how long the process of rehabilitation of some of the peripheral neighbourhoods is taking, and others still lament the destruction of historic buildings in the historic centre and of industrial buildings throughout the city. See Calavita, "Viewpoint." The point is that, all in all, Barcelona has done an excellent job in balancing economic growth and quality of life.

51 Barcelona City Council (1999) "RIBA Awards Gold Medal to Barcelona," *Barcelona Bulletin*, No. 162, July.

Chapter 4

Governing Barcelona

Pasqual Maragall

Now that I am leaving, after almost 15 years as Mayor of Barcelona, I have been thinking back to October of 1985, when I visited Leningrad for the second time, for the ceremony of twinning that city with Barcelona. Of the three or four days that I was there, I spent one day ill in bed, watching the clouds rush past over the Neva, carried by the wind towards the west.

That one unplanned day of inactivity was not a complete waste of time; I began a long letter to the Prime Minister of Spain and the President of the Autonomous Government of Catalonia, a letter that went unfinished.

Two years later, I sat down once again to that same letter. Once again, it went unfinished. But even so, it was not an exercise in futility. Much has happened since then. Barcelona has realized the dream that had gone unacknowledged for so long: becoming, albeit briefly, the capital of the world, a milestone that was accomplished thanks to our working as we had never worked before.

Since then, there have been several elections, with their accompanying emotions, debates and new governments. Companions have come and gone, some of them, sadly, forever. 1991 was a good time to try once again to address the President of the Catalan government and the Prime Minister of Spain, because we were setting out on a decisive year and a full four-year term of office in the municipal government (1991–95). It was a time to present our plans, our situation at the time, what we needed and what I believe we had to offer to Catalonia and to Spain.

In 1992 we spent a few days with the González family at Coto de Doñana, and I presented Spain's Prime Minister with the original version of my letter. But now, as I am preparing to leave my office, one that has been a constant source of fulfilment for me, I believe that the time has come to write the final version. This will be it.

I should like to discuss many different subjects, among them the 1992 Olympics and what they meant to the city. But I should also like to express to you the meaning of Barcelona, of the Barcelona that we all would like to be part of. I will also write about the Municipal Charter or Special

Law on the City and the Barcelona 2000 Strategic Plan. And, of course, about money; one of the subjects that inevitably arises at all of our meetings. And the quality of life in large cities, that eternally unfinished business that the political class — I am tempted to say the establishment — have yet to resolve for the citizens of this country.

And I should like first and foremost to make a small confession: my sincere admiration for the President of the Autonomous Government of Catalonia, and the mixture of respect and solidarity that has always guided my attitude towards him, over and above whatever concurrences or discrepancies may arise between us from time to time.

With regard to the previous Prime Minister of Spain, to whom I addressed and address this letter, I have come to regard him as the person who embodies more than anyone else all of the ambition and wariness of my generation. I see myself cordially and intellectually reflected in him, as I believe do many thousands of Barcelona's citizens of my generation and of other generations as well. It has not been easy for the Barcelona of today, a proud and demanding city, nor for the historical Barcelona, a city so repeatedly injured and offended, but also so often victorious, to command any sort of respect or recognition from south of the Ebro.

Aside from any differences of opinion, all Catalans look up to the President of our Autonomous Government, to whom this letter is also addressed, as the voice of our nation. He acts as the personification of Catalonia, of our intransigence and flexibility. He is the passionate interpreter of our moment in history and has presided over the emergence of the Catalan Autonomous Government as a fully mature institution, respected by all and envied by many. And we should not forget how all of this was made possible by the honour maintained for so long by President Tarradellas, culminating in his most dignified return.

Even those who, like myself, have suffered the stings of the stance taken at times by President Pujol, in keeping with his own innate sense of the gravity of his historical task, have come to recognize that he always takes that stance for a reason.

The Olympic Games

As you are well aware, the organization of the Olympic Games centred on two equally vital elements: urban infrastructure and the working of the Games themselves. I can assure you that both of those elements were handled with great care, as evidenced by the huge success that we achieved. The most difficult time was 1987–89, after our city was chosen in Lausanne at the end of 1986, with the support of your personal presence, and the arduous transformation of the Candidature Committee into the Organizing Committee, until the initial results of our constructive undertaking began making their presence felt in public, on the streets. It was, as you will recall, a rather dramatic beginning.

4.1
Lamp standards at the Olympic port

An Olympic city is neither invented nor created by decree: it is built up slowly on the underpinnings of an essential vocation. Then, gradually the city learns to be Olympic.

From today's perspective, all these sufferings seem well worth it. Above all, the experience shows the correctness of keeping to the planned direction without great changes and keeping faith with the initial team, although, as inevitably occurs, some had to pay for all, the unjust price of politics. It was important to be guided by cold reason, rather than by opinions generated by anxiety. For infrastructures, the working group formed to monitor progress was very valuable. Its fortnightly reports, based on external or senior management review, recounted the difficulties of each of the 300 projects: the airport, the ring road sections, the sports installations, the rowing lakes, the hotels, the National Palace …

As is well known, and as is accepted by public opinion, the overall result was very positive (Figure 4.1). The "pinch points", in the language of construction engineering, were successively the moving of the railway lines from Poblenou, the airport, the placa de les Glories, the sports installations for the tests of the summer of 1991, the Joan Carles I and Arts hotels which needed a reasonable period of use before hosting heads of state and members of the Olympic family; the National Palace of Montjuic (Figure 4.2),

4.2
Olympics era development on Montjuïc

GOVERNING BARCELONA   **67**

the site for the inaugural social event, where we received those invited from around the world and where Catalonia could again show its Romanesque and Gothic art; and the Trinitat intersection, the Gordian knot where the two ring roads and the motorway from France joined. And there was also the stretch of the coastal ring road, which was soon to link that road with the Diagonal and go down to the river Llobregat and then cross it, in front of the airport.

Throughout 1991, a year that now seems long passed, we had the opportunity to review the situation of one of the projects that I found the most exciting: the Palau Nacional. This old hulk, a remnant of the 1929 Universal Exposition, was to be transformed into Spain's second great museum, surpassed only by the Prado. It was an exciting project because it is the largest building in Barcelona and one of the largest museums in the world. The Palau Nacional houses what is beyond question the most complete collection of medieval art and particularly Romanesque art. The republican tradition and the law granting autonomy had made it Catalonia's great art museum.

In the end, the miracle was worked, with everyone pulling together. During the summer of 1992, this symbolic edifice stood out on Montjuïc, along with the Stadium, as the great monument to the Olympics, Catalonia's calling card and the best planned of our investments in our future draw as a quality destination for tourism. This summer, we inaugurated the Gothic section. It only remains to wish that, ten years after the works began, the project continues to its conclusion without delays.

All this work was possible because the operation was in good hands. Pere Durán and Gae Aulenti, with Enric Steegman, the Italian construction company Edil Fornaciai and the indefatigable Josep Anton Acebillo should be mentioned here. In this way it was possible to guarantee the no less important economic balance of this vast operation. A not insignificant part of the success lay in reinforcing international and security relations, as well as those with the media, which at first I imagined would not give us an especially easy time. This was not the case. Information multiplied about Barcelona around the world, and did not present us at all badly, one of the keys which contributed to our success and the positive image the city gave to the world.

In the months leading up to 25 July 1992, I took a personal interest in all of these aspects, even though we were convinced that the works were advancing as planned, and despite my certainty that the operation would work perfectly as far as information technology, telecommunications, sports competitions and logistics were concerned. The team, I repeat, was first rate and worked very well together, under the excellent leadership of Josep Miguel Abad; I did not repent of having defended him at moments of crisis. He returned the confidence many times over.

I wanted to assure the head of the Spanish government that the Mayoral Presidency of COOB 92 (the Barcelona Olympic Committee) understood the spirit of his words: keeping security as a priority, and working politically and in detail to ensure an image of the country which corresponded to the effort made and the enormous opportunity we had before us.

The President of the Generalitat had an important role in the months before the Olympic Games, and I asked for specific guidance on what was necessary in that respect.

Catalonia, as well, could not afford to pass up such a historical opportunity to do what it had been prevented from doing in the past; to manifest itself through action.

If Spain justifiably wished to demonstrate that it had changed as much as the rest of the world suspected, Catalonia's intent was to show itself to the world as a fully fledged nation. And Spain could not turn a blind eye, it had to help. In this sense, then, the President of the Autonomous Government played a crucial role.

He, more than anyone, had to guarantee and be responsible for the tuning of the plural messages which, as could be expected, would be produced. This he did irreproachably, with the exception of some particular episodes which did not affect the overall trajectory. The President knows that in those months before the Games I put myself in his hands to deal with issues whenever needed. No one was more responsible than ourselves in the development of the whole project. For my part, I knew that I had to be ready for the sacrifices and major efforts that were necessary.

For example, it would not have been good if people had got the impression that we were changing things because of the events underway in the ex Soviet Union. Some of the proposals that I made in small and discrete circles about elements of plurality in the opening ceremony could have been interpreted as showing excessive sensitivity to external events. I had to discuss this with Jordi Pujol himself and with the president of the International Olympic Committee, Joan Antoni Samaranch.

Furthermore, the international situation during the preceding years was developing in such a way that the Barcelona Olympics, which had from the very start deliberately projected an image of festivity and international *rapprochement* on multiple levels, could actually provide an opportunity that worldwide public opinion had been anxiously awaiting: the opportunity to see whether the huge advances that had recently been made could win out over the formidable economic, political and ideological difficulties and divisions opposing those advances.

But it is certain that this made things not easier but more difficult, in security, in protocol and in the messages to be conveyed, as well as in symbolism, languages and gestures.

We worked passionately for everything to work well, and it did work well.

Barcelona, however, is not content with having made a great leap forward, as it had already done in both 1888 and 1929. It is aiming for a permanent place among the world's great cities. It wants to stay at the top and not lose any ground.

The Olympic project was a magnificent spur to the energies of the city. It channelled external resources at a level which had always been desired but never obtained in the past 60 years, a flow which we would want to be continuous, not sporadic.

The Olympic holding company and the other investments made directly by central government and/or the Catalan Autonomous Government made it possible to mobilize, along with private and local investments, funds amounting to some 750,000 million pesetas, or as much as 900,000 million pesetas, according to some sources, over the relatively short period of four or five years. They allowed the city and metropolitan area to be endowed with good communications, telecommunications, hotels, alternative tertiary centres, storm drains, stadiums and other sporting installations, a marina, airport, main roads, and also the reform, whether completed or begun, of major cultural structures such as the National Palace, the Museum of Modern Art, the convent dels Angels, the Casa de la Caritat, the monastery of Pedralbes, the National Theatre, the Auditorium, the Archive of the Kingdom of Aragon, the Museum of Contemporary Art of Barcelona (MACBA), the new Lliure theatre, the Liceu, the renovation of the Palau de la Musica, the Museum of the History of the City, the botanical gardens …

A European metropolitan system

There is still much to do: particularly finishing the momentum in cultural centres, ensuring the supply of clean water and improving the river Besòs area, and creating the logistic activities area of the river Llobregat, to which I will refer later.

Already structured in its physical and social reality as a metropolitan area of three to four million inhabitants, Barcelona has to keep alive the sources of its qualitative growth and competitiveness as a city.

The population of the central portion of the metropolitan area, the municipality of Barcelona, is decreasing, in the sense that fewer people actually sleep there. The metropolitan area as a whole is not growing, and neither is Catalonia. There is therefore a movement of population towards the periphery of the urban area, and a new demographic balance between the metropolitan area and Catalonia, after the 25 years of concentration (1950–75). The new situation presents problems of quality of life and equity amongst different sectors of the population, but no major worries for territorial strategies. It is not a matter of aiming to make either Barcelona municipality or the metropolitan area grow. It is a matter of attracting interests, investments, technology, academic and scientific activities, highly educated people, high level cultural activities. This needs to be done so that it does not reduce the factors of attraction of the urban centres of Catalonia, in fact increasing these, so as to increase the connection of the metropolitan-urban system of Spain within the larger system of cities of Europe.

The major decisions taken in recent years affecting Barcelona have benefited the whole of Catalonia and have provided Spain with an incipient metropolitan system that can compete with Europe's urban networks, with considerable positive results for the peninsula as a whole. But some things are still missing, as I have said.

We lack water and space. Up to 2000, in less than ten years we will have invested 100,000 million pesetas in water treatment and new sources of water, and we will have to take decisions about new external sources, even though the much criticized absorption of ecological costs in water bills has allowed adequate financing and a clear holding down of consumption.

Barcelona has presented the possibility of creating a double network of water distribution, one for drinking and one recycled, and wants to contribute to the conversion of the rivers Besòs and Llobregat into natural and normal landscapes which are valuable and clean. The two goals are complementary, as without water treatment, we cannot get dignified landscapes or enough water.

My position on the metropolitan areas is well known. The dispersal of its governance amongst various sectoral and territorial organizations was a political and strategic mistake, certainly without large electoral effects in the short term, and dictated by a calculation of territorial rebalancing in Catalonia which now has no sense, given the spontaneous trends underway. The most outstanding eras of the history of Catalonia have been linked to a politically and economically strong Barcelona. It is not true that Catalonia's economy requires, demands or can even permit a weak Barcelona. Nor is it true that Barcelona has gained its strength from having stepped into the gap left by an inconsistent and insufficient Catalan power, a power that is not well received by the rest of the peninsula. Barcelona has made that substitution. And, in a way, all of Catalonia has survived to some extent thanks to this fact. It is true that Barcelona could not have done this if an underlying feeling of nationality had not existed, expressing itself through names, flags and symbols of the city.

But it is wrong to think now, with the recovery of the path of the rich, full, autonomous and confident Catalonia, that Barcelona has to renounce its strength, which continues to be the best guarantee of the freedom of Catalonia. This is how it has been in the best moments of our history, not just in the difficult ones.

It is true that there is a subtle game of meanings (Spain, Catalonia, Barcelona). And it is still true that the strength of Barcelona (what you, the President of Catalonia, once called the *force de frappe* of Catalonia) would not have the consistency and depth it has without the existence of a very rich and compressed urban network, of which Barcelona is just the headline. Barcelona has to collaborate with this network so that it is respected and strengthened as the mesh of the country, because from this it obtains the demand, the roots, the stability and the historical base of its existence as the capital.

At the same time, Barcelona would not be what it is (nor would Catalonia be what it is), in spite of the obstacles, ignorance and aggressions from outside, if it had not been inside a larger market, country and state, not always easy to live with or benevolent, but always present as an economic environment, as a cultural and political challenge, as a source of labour and talents for the construction of our own national and urban life.

Spain would not be Spain without Barcelona, just as it would not be without Catalonia. This means that with the normalization of our inter-territorial relations within the peninsula, we must leave room for a strategy for Barcelona — and for Catalonia — worked out by Spain, by all of the different communities that make up Spain, by a State of which we are an important part, sometimes a decisive one, as the doorway to Europe and a driving economic, cultural and technical force.

Any other approach, though perfectly respectable, would be mistaken in allowing us to settle for less than we can be, with less than we are.

At the end of 1909, Joan Maragall, in an article dedicated to the Empordà, opened up a line of thinking in which the meanings Barcelona, Catalonia, Spain and Europe took the form that we give them, approximately, today.

We are European because we are Catalan, like the continent's other nationalities. Catalonia is the heart and soul of our European character. The best way for us to be good Spaniards is to be good Catalans. And the time will come when we will not even have to say that we are Catalans. It will be sufficient to say that we are from the Empordà or from Barcelona. This is the Maragall message. The essence is to be found in the part rather than in the whole, in the seed rather than in the fruit.

Our city received the world in this framework of certainties, conscious of being the capital of a unique and unrepeatable national culture, and knowing that this culture would not have all the value it could have if it was not set within a federal construction, respectful of modern and plural Spain. It was also conscious of being the temporary capital of the hopes for fraternity and sporting and peaceful competition of thousands of millions of men and women around the world.

What is good for Barcelona is good for Catalonia, and what is good for Catalonia is good for Spain. These are truths that are only lacking one condition in order to be wholly and profoundly true: that they can be expressed in the context of a growing confidence that the sacrifices that they imply can be admitted without regret, as the normal and reasonable price for the exchange of sentiments and reciprocal contributions required by coexistence within a project shared by all of the communities of Spain, looking towards European unity, pacification of the Mediterranean region and ties of brotherhood with Latin America.

Barcelona offered to Catalonia and Spain the possibility of playing (intensely for some days and more modestly and constantly afterwards) the role of dynamo for projects supporting coexistence.

World centre of peaceful competition, vanguard of Iberian autonomy, a northern reference point for southern Europe, motor of dynamism for the cities of Europe, the largest city of the Mediterranean coast of Europe, European cultural capital: this is what we want to be, and this is what we are in some senses already, with your help.

The Municipal Charter, or Special Law of the city

In 1904, Cambó asked Alfonso XII for a law for the city. Jaume I had given such a law in 1284 with the Recognoverum Proceres, the content of which was later developed in the Ordinacions d'En Sanctacília. Philip V revoked these in the Decret de Nova Planta, which pushed back the territory of the city (which had stretched from Mongat to Castelldefels), returning Barcelona to the area within the walls, plus an arc defined by the distance of a cannon shot from the walls, the current Eixample.

The city revived and recovered the territory before the legal situation recognized this. Less than two centuries had passed before all the municipalities created by the Nova Planta decree on the plain of Barcelona, with the liberties of some parishes within the walls, had been annexed by the central municipality. Sant Martí de Provençals, Sant Joan d'Horta, Sants, Sant Vicenç de Sarrià and Sant Gervasi, Gràcia, Sant Andreu de Palomar, les Corts, all became part of Barcelona. The municipalities or parishes beyond the rivers remained outside. From 1860, physically the new large city was a fact, with the demolition of the walls and the construction of the Eixample.

Cambó's demand was therefore the crowning, the admission of an undeniable fact. For some years poets had sung of the city extending from river to river, defended by the mountains of the Garraf, Sant Pere Màrtir and Montgat, and the "breasts of Catalonia, Montseny and Montserrat". Now we ask, respectfully, after more than ten years of preparatory work, and when the few gaps left in urban development are being filled and the new networks of transport routes are completed, that the juridical particularities of our situation are recognized in three fields.

In the first place, we want to address the special relationship between the central municipality and those surrounding it, all united in a single regional agglomeration of more than four million inhabitants, needing a common administration of major infrastructures and united territorial planning, as well as the specific collaboration of the municipalities which form the continuous built up central area of the agglomeration and which should administer together the lesser planning matters, and perhaps a series of public services, now cut up by the existence of so many local frontiers.

Second, we must address the field of collaboration between different institutions in managing infrastructures and services of major significance, the competence of one or other body, basically of the state: the port, the airport, the central markets, the trade fair, the Palace of

Congresses, the Zona Franca, the logistics activity area, the technology park. For example, for the airport to work well the Generalitat, Barcelona council, the employers' federation, the council of El Prat should all be represented; this requires in advance the transformation of the national airports agency (AENA) into a holding company, at least for the airports of Madrid, Barcelona and Palma.

We can add to this, although they have already been mentioned and have a different character, the cultural bodies, such as the Liceu, the Auditorium, the National Art Museum of Catalonia and the MACBA, in which the public administrations present in the city have formed slowly the most appropriate systems of collaboration. In this area we can observe with optimism the current situation, which will make Barcelona what it has always had the potential to be, a European cultural capital of the first order.

Some examples of the cultural boom are the CCCB, the National Theatre, the Theatre Institute, the Palau de la Música, the Born library, the Library of Catalonia, the Institute for Catalan Studies, the Archives of the Kingdom of Aragón, the convent dels Angels, the Botanical Gardens, the old and new museums and cultural centres — such as the Barbier-Müeller, the Catalan Museum of History, the Science Museum and the Casaramona — the Aquarium and the proliferation of private theatres and galleries.

Still in the field of the articulation of the major Barcelona or metropolitan bodies, the council (without wishing to exaggerate its claims) sees the representation of society in the universities or saving banks as weak. Perhaps the special law for Barcelona could do something on this issue. The lawyer Puig Salellas has proposed an intelligent formula on the first matter. For the second, not even the Defender of the People Ruiz Giménez dared to oppose a challenge, which he knew to be well founded, against the Catalan law of savings banks which virtually suppressed municipal representation in the savings banks, as this occurs in the rest of Spain.

The council of Barcelona, with that of Sabadell, gave land for, and in fact led the process of forming, the Autonoma University, the governing body of which was chaired by the deputy mayor. The Diputació (provincial government) gave premises (next to the council house) and money for the foundation of the savings bank of Barcelona (la Caixa).

Finally, in terms of administration and finances, Barcelona has always been an innovator. Why not recognize that Barcelona is in the vanguard of a series of processes that need to be supported by law, which can contribute modestly to the solution of the problems of large metropolitan cities?

I will return to this at the end of these letters. In Barcelona, the legal framework should be the Spanish constitution, the Statute of Catalonia and our Special Law.

We understand that higher levels of government see this historic Barcelonan request as a plea for special treatment. But we also understand that if this special situation has been recognized over the centuries, even in non-democratic periods, reality and the arguments coming from deep within society in the end impose themselves on temporary arguments.

Barcelona cannot go on making such an outstanding contribution to the modern configuration of a democratic Spain and an autonomous Catalonia unless it has at its disposal the legislative instrument for shaping its destiny that it aspires to, deserves and needs.

In short: to govern relations within the complex of the metropolitan area and the historical centre, to achieve a stable balance in relations between the capital and the rest of the country, to ensure the participation of the city and the Autonomous Government of Catalonia, along with the central government and society in general, in the management of the principal institutions and technical, economic, cultural and other structures, to create the essential local justice system and a system for immediate justice, to provide the city with the means to deal with issues concerning the environment, transportation and housing, and the financing and provision of social services and education. This is the goal of Barcelona's Municipal Charter, the body of rules and regulations required to allow the city to continue its progress.

The Strategic Plan Barcelona 2000

Two hundred bodies have worked for years to produce a stimulating and credible future scenario. You know well the conclusions of the work. In the first Plan, fourteen working groups, led by a convenor, worked to define the objectives in fields where there is no clear administration with competences. In total, 54 objectives were formulated, and half of these have been achieved — a good result (Figure 4.3).

4.3
Publicity information of the Conventions Centre under construction in 2004 Forum area (2002) achieving a Strategic Plan objective

GOVERNING BARCELONA 75

The three main areas of agreement were the organization of a macro-region covering the northern part of Southern Europe as a precondition for progress, the priority of services for people and quality of life as the specific framework for the philosophy shared by all of the citizens of Barcelona, and a focus on providing services for business.

If we remember that in the 1920s Barcelona was, in the eyes of Europe, the city of bombs, that a novel with the title *When they killed in the streets* could be successful, that the deepest hatreds and the most extreme intolerance were present in the 1930s and 1940s, that the current President of the Generalitat was mistreated by the police in the 1960s for having dared to criticize the militant anti-Catalanism of a newspaper editor, and that in the early 1970s we lived in a state of emergency, with meetings limited to 20 people, political executions … who can deny that it is not a small miracle to have formed a consensus between 200 bodies about our future, including trade unions and employers organizations, business and the universities? Is that not the best summary of the political transformations of Spain, Catalonia and Barcelona in the past 15 years?

The King himself, who was the first to be presented with the conclusions reached under the initial Plan, on 25 May 1990, is an enthusiastic supporter and the similar actions undertaken by Madrid and Seville were stimulated by the favourable reception of those conclusions in influential circles. The OECD has had a detailed account of the Strategic Plan and recommended it as a model for its member countries. The Ibero–American Centre of Urban Strategic Development (CIDEU) has advised around 20 cities of the new world following the Barcelona model.

The result of the spirit of the first strategic plan meetings and of the support of the minister of economy of the Generalitat, Macià Alavedra, was the creation of the first joint initiative of the Barcelona financial community, European Financial Centre, first chaired by Claudi Boada and then, alternating, by the director generals or presidents of the two large savings banks, the Banca Catalana and the Bank of Sabadell. We do not want to be on the edge of financial decisions in Europe. Our economic potential both justifies and demands our participation. We are prepared to accept financial specialization and modesty, but not disappearance. We know that there is a place for us in this context and we intend to do everything in our power to occupy that place. The Futures Market, piloted by Basáñez i Oller, makes us believe that we can achieve this, as well as the revival of the Stock Exchange, led by Joan Hortalà.

In the same way that we will strive to overcome our weaknesses, we will also strive to take advantage of our unique nature. Barcelona's density makes it a city on a human scale, a welcoming city that is loved by its inhabitants and by its visitors, and it exempts us from the approaches characteristic of the large national capitals, it forces us to organize the use of available space and perhaps endows us with an extreme sensitivity to detail and form: Barcelona is a conjunction of metallic forms, concrete twisted into the unlikeliest shapes, carved wood, walls decorated in Sgraffito, convoluted glass, shining facades, narrative balustrades, worked stone and re-worked stones, jewels, stucco, retables, surprising nooks; yet each square foot tends to be crafted in this city that some would characterize as subject to the tyranny of the square foot, to the miserly lack of space and to the dominance of an aesthetic decried by Unamuno and praised by Cervantes.

The year 2000 and the twenty-first century serve as a framework for a new pact with our surroundings. We need a little more ambition in the size of structures, without losing our humanism or prudence. There can be some tall buildings, but not many, enough for a generation which, in spite of appearances, wants to preserve above all the spirit of a well-proportioned, just and contained city. In this context, there are two aspects which I always mention, the cabling of the city and the treatment of the subsoil.

It is a dense city, a congested city, an expensive city: three aspects that are so interlinked that no one is quite able to organize them into an equation of cause and effect. Economists like to tell us that land is expensive because of its scarcity and that the city is congested because of its density.

But Barcelona has worked pretty well historically as a matrix of flows and movements. The rationalist Eixample of the nineteenth century is an international model of the optimization of the number of exchanges per unit of land and time. Will it resist well enough the current siege by motor vehicles? We will discuss this later, and will keep talking about it.

In any case, to maintain and increase its attractiveness, Barcelona can take advantage of its density to achieve low cost communications per capita. Our city piles up many inhabitants and businesses on each hectare of land. If the land is equipped, cabled, connected to communications networks, this is cheap for each user — hence the importance of fibre optic cabling.

While I am putting the final touches to this letter, we are about to obtain for Cable and Television of Barcelona (in which our own Catalana d'Iniciatives is a participant, and which began this body in 1986) the contract for the three Catalan cable areas and, we hope, for the business led by ENDESA, the winning of Retevision, which will be based in Barcelona. These two results stem from years of work and will place us where we want to be in this decisive field. Ernest Maragall has had a great deal to do with this achievement.

In the same way, and in order to regulate an extraordinarily congested subsoil, we are pioneers of underground cartography, the creation of service galleries and the collaboration between local authorities and public service companies.

Another outcome of the Strategic Plan is the C-6 network of the cities of the northern part of Southern Europe (Montpelier, Toulouse, Zaragoza, Valencia, Palma de Mallorca and Barcelona). This group of cities and their corresponding areas of influence are home to fifteen million inhabitants — 5% of the population of the European Community in 8% of its territory. Their unemployment rate is higher than the European average, but their rate of economic growth is also above average.

It does not seem as if municipal political changes have to damage the cohesion of the Barcelona region. This region brings together all the circumstances required to form one of the balancing territorial complexes of a Europe which tilts much towards the north, and also since 1989 towards the east.

In the same way that the Strategic Plan logically favours European gauge over the trend to high speed, and has always argued for the network's southern connections (Barcelona–Madrid–Valencia), we also feel that we have a right to expect intelligent understanding of the benefits for the whole peninsula, as well as for the French *Midi*, of a strong border region, straddling the Pyrenees which will cease being a barrier and become instead a crossroad, as the Alps have been for centuries in Central Europe. The Trans-Pyrenees Local Administration Committee (Seu d'Urgell, Arrège, Andorra, Barcelona and Toulouse) has joined the C-6. The dream of our seers is coming true: the Pyrenees will matter. The Cadí and Puymorens tunnels are bringing it closer to reality. The Pyrenees Citizens' Charter is its expression.

One constant source of dissatisfaction has been air connections and they are now on track to a very different future, thanks to the new terminal and plans for a third runway and new train connections. The change of management model, implied in the previous section, and of course the change of attitude of our leading company and of the many small companies operating in Barcelona in the treatment of the rising demand for international flights from Barcelona, should lead to the formation of a dual peninsular airport space and to better use of the advantages of Barcelona's position. For the formation of this airport hub, for hangers and air services (today a large plane cannot be checked over, cannot sleep at Barcelona), we have to sacrifice other significant projects which are not so strategic in the airport zone, such as the large trade fair complex which was proposed some time ago, or similarly large projects.

An airport serving transoceanic fights is not viable in Southern Europe without a market of fifteen million inhabitants to support it. But even this virtual market could be insufficient if comparative costs and quality of service are not up to the required standards.

The Port-Airport-Mercabarna-Zona Franca-Trade Fair-Logistics Zone complex could at last, by the turn of the century, become a modest competitor for the large distribution centres of the European Atlantic coast (Amsterdam, Rotterdam, Antwerp), through which the largest part of the merchandise, information and passengers from the Americas and the Pacific enter the continent.

We will dedicate a considerable part of our strategic efforts to the strengthening of these activities and their territorial coherence.

But there is no doubt that the capacity of attraction, the interest and the quality of a city depend basically, in the last analysis, on the quality of life of the persons who live and work there. We dedicate the next but one section to this point, which the strategic plan sees as central. But first I must address the issue of money.

Finances

[The section on the financing of the city has been abbreviated here.]

In 1987, I said to you: this is a particularly critical but also promising moment. Thanks to gradual progress on the basis of studies and negotiations with former governments, the central government assumed a portion of Barcelona's debt amounting to some 100,000 million pesetas, after the Socialist party came into power in 1982. […]

Barcelona's City Council has shown itself to be a good administrator of the resources at its disposal: it has increased the level of services while the number of employees has decreased since April 1979, it has instilled discipline in the ranks of its employees, it has improved the administration's image, it has a better credit rating than it had before General Franco slashed the face value of the bonds issued to finance the Universal Exposition, it borrows on the Japanese money markets with no need of any guarantee on the part of the Kingdom of Spain and under the same conditions, and it has been a key contributor to finding solutions for ten or twelve black spots that were the real basis of and force behind our financial crisis. […]

Barcelona's City Council has gone from being an agent of crisis to being a provider of solutions, contributing to the resolution of complicated problems linked to banking difficulties (Montigalà tunnels and industrial estate) or to the historical inefficiency of state-owned companies (SEAT, MTM, ENASA) or of companies owned jointly by the central and local governments or by local government itself (Mercabarna, IMAS, Pompas Fúnebres, etc.). This fact is generally known to suppliers, contractors and financial intermediaries. You, as heads of governments, can verify this. […]

The metropolitan question is far from trivial.

It is in fact at the heart of the most serious of all of the problems that the past has seen and that the future holds in store.

The 26 municipalities that surround Barcelona, with 1,300,000 inhabitants, together make up a large part of the negative factors that mar the brilliant image projected at times by the city centre. […] I should like now to discuss the costs of being a capital city.

Barcelona has come to substitute for the central government in many areas, either as a stop-gap or as a result of its own power.

In some cases, it is now substituting for the Catalan Autonomous Government, owing to the transfer of the corresponding services. I would say that is the most common case.

Healthcare, basic and professional education, museums, cultural and financial institutions, as I have already mentioned (Palau, Liceu, etc.), the Conservatory, and the City Orchestra are the most noteworthy characteristic services in this sense.

Who should pay for these services? Or, rather, through which levels of government should citizens — who, in the end, are always the ones who pay — have to finance these services?

Where these services have been transferred, it would seem obvious that the funds should be provided through the Autonomous Government. But how can this be done if the actual cost of these services was not calculated when the transfers were made? […]

The situation with regard to museums became even more paradoxical when the entry charge was eliminated from national museums. We could not afford to do the same. Now, foreign visitors to Barcelona may come away with the impression that Barcelona is less generous than other cities, when Barcelona's citizens are the only ones who pay for their own museums. […]

[…] One year before the Olympic Games, Barcelona was carrying a long-term debt of 100,000 million pesetas, the logical reverse side of our common effort to create infrastructure, as well as a short-term debt of a further 100,000 million pesetas, the algebraic sum of the annual deficits resulting from our payment of services for third parties and from delays in the implementation of already agreed solutions: the write-off of debt up to 31 December 1982 and the 1985–90 agreement programme.

[…] Barcelona has never asked for the return of that contribution. Nor has it asked that anyone else pay its debt for it. We can pay these debts and we are paying them. What Barcelona asks is that it be compensated once and for all for the expenses that it has incurred on behalf of the central government or the Autonomous Government.

Six years ago, the city made this request so that it could confidently strive for success in the decisive year of 1992 and its reasonably active aftermath with the proposal that 20,000 million pesetas be invested yearly — the same amount as during the years prior to that turning point, without counting investments in the Olympics — with higher amounts for maintenance of the increased infrastructure. The city requested this because it was unreasonable, even while acknowledging all of the efforts that were made on our behalf, that our tax burden should be heavier than it was in the five years prior to 1992 — already heavier than in most of Spain's large cities and much heavier than in Madrid, to cite the usual example.

In order to manage this debt effectively we required and continue to require cooperation of the Finance Ministry and Autonomous Government's Department of the Economy. […]

We accept, as is generally accepted, that the goal for the year 2000 of sharing out net public spending between the three administrations in the proportion of 50/25/25, equivalent to an equal sharing of expenses between the central government and the regional governments, and also between the Autonomous Government and local authorities, cannot be met without a simultaneous or subsequent transfer of powers and obligations to the local sector. [...]

The quality of life in the large city

The other motive I have for writing to you, more important to me even than the question of money, is what we could call the social objective of our governments: the quality of life.

There is a dominant impression that the governing bodies are not giving responses to the new challenges, ever more qualitative and difficult to pin down, which are especially present in major cities.

Spain, and Catalonia and Barcelona in particular, would receive a very high grade in the classical subjects of government from a jury formed by other countries and cities from around the world, except perhaps with regard to the question of terrorism, if indeed that question may be judged from the outside and does not in fact require, as I believe, a wise reserve on the part of international observers.

But in the same way that we may be held up as models of good government, we are not so worthy of praise with regard to the quality of life of the people we govern. We are told our cities, and particularly Barcelona, are noisy, frequently dirty, unsafe, with expensive housing and with slow moving and excessive traffic of private vehicles and therefore also of public vehicles, lacking in solidarity, unable to do away with abject poverty or to prevent the decay of entire neighbourhoods, powerless to deal with the waves of immigration that are the other face of the coin of our own progress, a seed bed for aggressiveness, in demographic decline: an endless list that our own media, our political parties — when they gather to hold their congresses — our neighbourhood associations, our own children ceaselessly recite to the administrators who do their best to avoid listening to them. It would seem then that we, who excel in pure political and infrastructure administration, are building a magnificent new country, or magnificently regenerating an old one, transforming our city beyond recognition … simply to make them a backdrop for our own persistent inability to make any real improvement. The best of stage settings and the worst performance, the same old play.

And perhaps in reality not much more can be done. At the most, the need we feel in our hearts of having contributed to objectives which really have to do with what counts for humanity (the improvement and dignity of our human and environmental surroundings) we can satisfy through what we manage in terms of collective pride: Spain is more respected, Catalonia freer, Barcelona more of a city than ever. Others tell us that. Our old people feel more protected, economically and in health. Young people do not say so, but we know they live in better conditions. But still — are we really progressing?

The deeper motive behind this letter is the conviction, which I am convinced that we must have in common, that we can make headway. That we are on the verge of making headway. That in the goals of our collective life, in the most demanding sense, in the sense most closely linked to our social morality we are making progress or we could make progress.

And if it has some sense that the mayor of a city dares to address himself to yourselves in this tone, it is because he thinks that he can contribute, from a particular perspective, to the adjustment of the norms of our collaboration, to this end. It is because from the Mayorship, one sees aspects which I think it is interesting to communicate to higher levels of government. I do not intend in these letters to get into the vexatious issue of inter-institutional relations, especially within Catalonia, which makes us uncomfortable, both the President of the Generalitat and myself. We both know what we have had to swallow and what we have had to be forceful on, to respond with dignity to the clear respective requirements of national and local pride. We both know that the results of this relation (whatever is the meaning that this merits in public) are, in this respect, enormously positive. And we know at the same time that we start from different principles which are probably not reconcilable in the course of our political activity, however long and continuing that may seem, though short in reality if one takes account of the time that history needs for really significant changes.

It is a matter here of collaborating in the ways that our governing teams know. Armet and Guitart in Culture, Trias and Clos in Health, De Nadal and Borrell in Finances, high buildings and traffic authorities, Raventós and Alavedra in Economic Development, Marta Mata, Rubacalba and Carme Laura Gil in Education, Eulàlia Vintró and those in charge of Social Welfare (a portfolio incidentally created in 1987 by the council and then by the Generalitat in 1988 and the State in 1989), Jordi Borja and those in charge of International Cooperation, all of these have shown the capacity to understand each other. Now we would have to widen and bring up to date this list, but I leave it as it was written some years ago because it gives the flavour of the time and because as time passes it is ever more just and necessary to remember old situations and names and positions past, forgotten.

Was our agreement necessary for this to happen? Without doubt, directly or indirectly. When tensions are created at the top, cooperation suffers at operational levels, although often the force of things, the interest and self-respect of those responsible are stronger than the climate at the top.

Nevertheless, there remain a good many areas where things are not working as well as we would like them to. Traffic and parking, housing and the decay of neighbourhoods, lack of safety and drug abuse are the clearest examples. Here, it is not that this is what we are told; we know that we have to do more. We have, it is true, made a great deal of progress since this was originally written, but the point must still be brought home.

Normally we pass on the responsibility as we can: you will not give me land, I will not construct housing; central government did not pass on the estates in good condition, I will not fill them; the autonomous government of Catalonia has not made the effort equivalent to that of the state in other regions who have not had such a transfer of powers and where the state acts directly; the state will not pass the money that you ask for; you ask me that the local police should have the character of judicial police, I tell you that the judges do not want that, that I have other problems more important than transferring money to the autonomous government's police, etc. And the triangular evasion is even more annoying than others (don't bother me now, I am dealing with this gentleman), a situation which logically happens often, in a system with three levels of government.

In these situations, I repeat, the cynicism of a certain part of opinion would be justified if we did not do something different and kept using up the time we have in extensive bilateral exchanges which do not solve the problems of the people, but just, in the best cases, solve the problems of the governors of the two levels involved.

Management of transport

Specifically, in the field of transport we have to recognize that mobility in the centre of cities, both public and private, does not pay the costs it generates, and should therefore increase its prices. The price of private mobility should rise more than that of public transport, because the first generates more congestion costs (queues, pollution, etc.). Finally, laws are not followed because fines are not paid (especially between different municipalities), and therefore local authorities, also lacking effective direct powers such as the withdrawal of driving licences or the immobilization of cars in situ, cannot ever, under these conditions, regulate traffic. I use superlatives to be clear, you must pardon my very definite tone.

Let me list the problems: low prices for mobility in the city centre, a scarce commodity, and even lower in the case of private mobility than public mobility, in comparison with what they should be, and as a result, few buses, buses that are poorly financed and adrift on a sea of private vehicles that are more polluting and suboptimal, as the economists like to say, relative prices for mobility governed more by macroeconomic considerations (the retail price index) than by any desire to improve traffic; constant pressure by automobile manufacturers to sell ever larger and more powerful vehicles that are destined to spend more and more time stuck in ever larger traffic jams; intimidation of the various governments by the magnitude of the importance of the automotive sector to GNP, exports and employment; the impossibility of building private car parks in the framework of an out-moded approach to town planning that does not recognize their status as useful facilities and, in general, makes no allowance for land dedicated to such a purpose. How few multi-storey car parks you will find in our cities flooded with cars, in the street and on the footpaths!

This is the situation, dramatically presented. We cannot do nothing. If we could do something, would we do it?

We have already done some things. We spent or committed much money to give the city an external ring road like those of Paris, New York or Madrid, and another transversal metro line, line 2. We built 50 public or leased car parks up to 1993, and another 35 more before the end of 1997. Barcelona is becoming perhaps the European city with most public car parks, especially residential ones. Why not private car parks? Because they are not profitable in comparison with building offices or housing, at current relative prices.

Noisy motorbikes have been being withdrawn for some years; sanctions by traffic police have increased, as have permissive experiments, for example with night-time parking in the street, in the centre of the city. The sirens of public ambulances have reduced their stridency and have a discontinuous sound, and in two tones, except for those that come from outside.

I fervently hope that in these questions not only the macroeconomic viewpoint is taken, but also the micropolitical one of congested cities, which need clear signs that extra-urban powers are aware of the problems, both in relative prices and in the resources available for financing the infrastructure and running of public transport, and in giving effective powers to the local councillors.

This area, I would say, is the main complaint and the main hope that a local authority can address to higher levels. At times, when the higher levels of government decide to take action and attack the problem of the quality of life in our cities — the González Transportation Plan, or the Pujol Housing Plan, for example — we are in two minds how to react. First we think: at long last, the increase in funding that we have been demanding for so long. Then we think: how many errors are going to be committed as a result of the distant macro or national approach to problems that we know inside out at the local level?

Although the problems posed by the quality of public spaces, traffic, noise, congestion, etc. may seem insurmountable, they can be dealt with. Not radically, perhaps, but substantially. People do not really expect much more than that.

Trust us, and we will make clear progress in this area.

I believe that much the same may be said with regard to housing and the decay of neighbourhoods.

We spent almost six years sparring on this matter, in spite of the fact that those responsible in this area were capable of transferring the municipal housing stock to the occupants and eliminating within ten years and without traumas the shanty town of la Perrona; what a lot we owe to Xavier Valls in this field!

We zoned land for public facilities in the old centres. We followed the slow and involved arrangements for the Areas of Integral Rehabilitation

(ARIs). We provided land to secure public or accessible housing, etc. The results were very poor.

In despair, I wrote directly to the President of the Generalitat, about the specific question of Ciutat Vella. It did not help much, and I completely understand, perhaps the same would have happened to me as to him, if I had had to understand the question of Ciutat Vella from a certain distance.

We decided on the creation of a company, jointly with private bodies interested in improving the district, to manage through the market, within the framework of the ARI, what we could not achieve with traditional administrative methods and the normal, hopelessly small public finances.

The results were considerable. We acquired discretely perhaps 300,000 m² of property, the equivalent of the Vila Olímpica. We demolished some blocks of houses, ensuring that the best factor for rehabilitation, sunlight, reached to the heart of Ciutat Vella. We cleaned up streets and public services, building neighbourhood facilities (not as many as the plans proposed), and attracted large institutions — universities, the MACBA, the Centre of Contemporary Culture, the Provincial Library, etc.

Ciutat Vella has changed. Young people go there. Artists and professionals want to set up shop there. The morale of the neighbourhood's three active groupings, neighbourhood associations, business associations and charitable organizations, has improved. The morale of the police force has also improved. I will comment later on the subject of the lack of safety. I have at times referred to the cross-country skier's technique — a town planning effort, then a public safety effort, and so on — to describe how public action should be taken in decayed neighbourhoods. But the key to rehabilitation is to be found in decentralization, proximity and familiarity with the territory. Barcelona has been fortunate enough to gather simultaneously the right formula, sufficient funds and the right people to manage rehabilitation efforts on the spot, in the districts themselves.

Does this mean that Ciutat Vella is now out of danger? Not necessarily. Pessimism still raises its head from time to time, although it is perhaps diminishing in strength. If the influx of immigrants with few resources increases at certain times, the process of assimilation can present problems.

Let's hope we show no faintness, even temporarily, in the rehabilitation process. In ten months we could lose three years. The key is in the teams involved, again. Those working are those who have to combine deep knowledge of the property market, a capacity for solidarity without dogmatism, collaboration with church authorities as well as with the national police or with shopkeepers, as well as a very clear idea of what can be and should be required of the judicial administration, as well as of health and educational services.

If our society was capable of rewarding the excellence in this field in some currency, perhaps other than with money, in such a diligent and categorical way as it rewards entrepreneurial talent, we would have here a select group of millionaires in this currency.

A housing and transport plan

The cities have learnt to watch out for *what others do*. Remember the periodic verbal incidents from poorly advised councillors, of various municipalities, in relation to the *export of beggars* and other phantoms, which, whatever optical illusions there are, always have a real basis. The collectivity of beggars, for example, has a great mobility and capacity to pass information between cities.

Excuse my digression. You may be asking, what the devil is the mayor of Barcelona wanting me to do about housing?

First of all, the general law is not always appropriate: the unit costs for construction and scales of income used in the law do not apply in Barcelona.

Second, you should remember that the municipality of Barcelona has reached its limits and is now full to overflowing: there is no more land for development.

Third, be aware that the only land available is underneath buildings that must be bought and emptied before demolishing them, to put it crudely, and that the process is therefore more expensive than in other circumstances.

Fourth, recognize that the real Barcelona, which is much larger than the municipality itself, has as much influence over the price of housing, investments in public transportation — as well as tunnels and ring roads — as over investments in housing itself.

Fifth and last, on the basis of the four preceding points, agree to a Housing Plan — and one for transportation — for Barcelona, one that is fully equipped to deal with the problem.

In any case, we should bear in mind that real prices in Barcelona are probably falling and will continue to do so. If you open the newspaper any day and take a look at the increasing amount of real estate advertising, you will see that there is competition. I do not like to admit it, but around 1992, behind most of that advertising was the municipality itself: in the Vila Olímpica, Montigalà, Vall d'Hebron, Eixample Marítim, Llars 93, Barna Rehabilitació, the porches of Fontserè and even the university flats of Bellaterra, another initiative that Xavier Valls got going. There is also Procivesa, the mixed company for Cituat Vella, in which I hope the Generalitat will end up collaborating. Today we should add: Diagonal-Mar, Front Marítim, Diagonal-Poblenou, etc.

This has been our policy: construct housing to keep prices down. We call accessible housing, housing which is sold with little or no profit margin; that is at cost. We do not think that extreme poverty can be solved by houses at below this price. That would have to be with subsidies or exemptions for individuals. Cheap housing does not make poor people rich, and it tends to impoverish the surroundings, reducing the values of the adjacent properties.

That is why we undertook the rehabilitation of 4 km of sea front with housing meant for the middle class and upper middle class. We did not want to have to deal with paying for the maintenance of 4 km of public housing out of the municipal budget: it would have been impossible.

The real estate sector in city centres, where the price of land, location and surroundings — and therefore the investment in facilities — are of such great importance, will see a very strong public sector presence, either operationally or through agreements.

We must offer public housing to all of those who are obliged by the city, either voluntarily or involuntarily, to leave home. And, in theory, to no one else. To a great extent, here, as in the area of finance, we are trying to implement solvent and sound processes and to deal in a dignified manner with the costs incurred in the past.

All of this will lead to the creation of a new class of urban administrators and specialized businesses, combining operational agility with the acceptance of prior goals set by the public sector.

The new Housing Plan is meant to support this process. Without the market, the Plan is worthless. On its own, the market destroys value just as easily as it creates it.

If I had written this last section of my letter, on public safety, some time ago, it would have been much more negative in tone.

Just look at the picture presented by the past few years: one third of the branches of "la Caixa" robbed in a year, a renewed increase in the index of victimization that we have been analysing since 1985, growing frustration on the part of local police, who are only able to keep a permanently high level of delinquency at bay by means of overtime and a certain degree of on the job heroism, repeated discouragement on the part of the most active judges and prosecutors in reaction to citizens' concern for this issue, the Bandrés motion for separation of hearings and sentencing, failure of the "rapid justice" operation on the Costa Brava owing to a lack of cooperation on the part of certain prosecutors, etc., a poorly motivated and understaffed national police force with poorly defined duties, the exasperatingly slow progress in implementing the plan for dignifying the police command and particularly the five main police stations, the also excessively slow progress being made towards reform of the prison situation, calling for movement of the Trinitat and Model prisons, a consular community that repeatedly submits its complaints every summer about the victimization of tourists, etc.

Nevertheless, the project for the Municipal Charter and the Olympic Games made the long-awaited miracle possible: the Judiciary Council, with the knowledge and approval of the Ministry and the Catalan judicial authorities, proposed the creation of four new duty courts and undertook thereby to speed up the judicial process and judge those accused of minor offences within ten days, for the duration of the Olympic Games and, of course, subsequently as well.

If Barcelona achieves this police-courts structuring, we will certainly benefit from a real improvement in levels of public safety and therefore in the efficiency of the justice system as perceived by our citizens. If not, on the other hand, we will find it very difficult to make any improvement. I do not say that the situation might become worse, because being aware as I am of the continued precariousness of these expectations for change, I feel obliged to express myself as if they might not exist and to cooperate, in this case, with a view that is more optimistic than the one that I actually hold, in order to maintain the impetus and the morale that will then be necessary. I believe that you understand quite clearly what I mean. Nevertheless, improvement has in fact been made. While a law-and-order based government like Mrs Thatcher's witnessed a 50% increase in criminality in ten years (1980–90), Barcelona improved its rate of victimization by 50% (1985–95).

We can summarize by saying that the famous neighbourhood police which exists since 1994 (more than two-thirds of the local police are assigned to districts) will not be effective until there is justice, not at neighbourhood, but at district level, or at superdistrict level.

Paradoxically, the proposal of the Council of Judicial Power responded apparently negatively to our idea for municipal justice, as was laid out in the Charter text at that stage. The Council told us: "There is no need for you to climb up to have municipal judicial power, we will come down to the district level". Or: "There is no need for you to take on a role in justice; we will decentralize our own organization and take on your concerns".

We were satisfied with that commitment, which seemed historic.

In reality we know that the Council is not against municipal justice, for a limited field of lesser offences — traffic accidents, neighbourhood quarrels, property problems in flat blocks, arbitrations, and perhaps even minor offences on the frontier of crime.

This is so amongst other reasons, because this *peacemaking justice* for the large city would solve a possible constitutional inequity in our basic laws, in treating in a different way the juridical protection of citizen security (professional justice) and that of the country dweller or inhabitant of small towns (non-professional).

This is an issue which remains open.

In any case, with a decentralized or de-concentrated justice system, whether it is municipal or not, it is obvious that the context will be created for the local police in large cities to act — first in practice, then later also in theory — as a judiciary police force. These changes can be brought about with minor modifications to the law, although these modifications will, of course, require attention to detail and a sense of proportion. There are certain modifications, such as the mention of graffiti in the Criminal Code, that we could well do without. You do not need to bring out the heavy artillery to kill a fly. However, if careful and sensible action is taken and if you have even the least confidence in my words here, believe me, it is on these minor changes that the quality of life depends to a very great extent. It depends on them, and not just on grand legislative, political or town planning programmes, which, even so, are indispensable.

You may rest assured that Barcelona will not abandon this double stance of attention and legitimate ambition. With the first, we will demand the instruments and resources that this great capital of a small nation requires to carry out its role in the context of Catalonia and Spain with dignity, autonomy and responsibility. With the second, we will make sure that relations between government and citizens are above all efficient and respectful, tolerant and decorous. After all, we must never forget that a city is first and foremost the people who live there.

Chapter 5

Ten points for an urban methodology

Oriol Bohigas

(In 1999, the Gold Medal of the Royal Institute of British Architects was given not to a person but a whole city, Barcelona. A key player in the rejuvenation of the city, Oriol Bohigas of MBM Arquitectes, made a speech in response at the award ceremony. It is a manifesto for good city-making, reproduced here in slightly edited form — editor, *Architectural Review*.)

Today our European cities are made not by the architects, or the engineers, or the urbanists, or the geographers or the economists, but by the body of the citizens, represented by their elected politicians. I would like to develop in ten points an urbanistic methodology, which can be deduced from the political reality of Barcelona.

The city as a political phenomenon

The city is a political phenomenon and as such it is loaded with ideology and with political praxis. In Barcelona it has been the continuity of a common ideology and programmes carried out by the city's three Socialist mayors, Serra, Maragall and Clos, and their collaborators, that has made the coherent transformation of the city possible.

The city as domain of the commonalty

These political and urban ideas are based on a radical statement: the city is the indispensable physical domain for the modern development of a coherent commonalty. It is not the place of the individual, but the place of the individuals who together make up a community. Very different from what a famous British politician said: that there was "no such thing as society", only individuals and the state. It is the relation between individuals that constantly weaves the threads of ideas and expanding information. The city offers the fullest guarantees for this information, for access to the product of that information and for the putting into effect of any socio-political programme based on that information. There can be no civilization without these three factors.

The new voices of technology have tended recently to say that the traditional city is going to find itself replaced by a series of telematic networks which will constitute a city without a site. This is an anthropological and ecological nonsense which stems from certain political ideas. It is a vision put forward by those who are opposed to giving priority to the collective and in favour of the privatization of the public domain.

Tensions and chance as instruments of information

When I say that the city provides us with certain irreplaceable instruments of information I mean the enriching presence of tensions and of chance. It is only with the potentially conflictive superimposition of singularities and differences and the unforeseen gifts of chance that civilization can progress, with the move from the structure of the tribe to the civilizing cohesion of the city.

The city is a centre of enriching conflicts which are only resolved in their affirmation as such in the coexistence of other conflicts with different origins. It does seem to me that the great error made by the urbanism of the Athens Charter was the attempt to cancel out these conflicts. To eliminate them instead of resolving them with the recognition of other conflicts. Urban expressways, the 7V of Le Corbusier, functional zoning, directional centres, the great shopping areas: these have not served to resolve problems but have instead destroyed the character and the function of many European cities.

The public space is the city

If we start out from the idea that the city is the physical domain for the modern development of the commonalty, we have to accept that in physical terms the city is the conjunction of its public spaces (colour plate 5). Public space is the city: here we have one of the basic principles of the urban theory of Barcelona's three Socialist mayors.

In order for urban space to fulfil its allotted role it has to resolve two questions: identity and legibility.

5.1
Plaça Allada i Vermell in Ciutat Vella (Old Town), created c. 1992

Identity

The identity of a public space is tied up with the physical and social identity of its wider setting (Figure 5.1). However, this identification is bound by limits of scale that are normally smaller than those of the city as a whole. This being so, if authentic collective identities are to be maintained and created it is necessary to understand the city not as a global unitary system but as a number of relatively autonomous small systems. In the case of the reconstruction of the existing city, these autonomous systems may coincide with the traditional neighbourhood make-up. I believe that this understanding of the city as the sum of its neighbourhoods or identifiable fragments has also been one of the basic criteria in the construction of Barcelona, with all its political significance and with the creation of the corresponding decentralized administrative instruments.

However, we are dealing here not simply with the identity of the neighbourhood, but with the particular representative identity of each fragment of the urban space; in other words, with the coherence of its form, its function, its image. The space of collective life must be not a residual space but a planned and meaningful space, designed in detail, to which the various public and private constructions must be subordinated. If this hierarchy is not established the city ceases to exist, as can be seen in so many suburbs and peripheral zones of European cities which have turned away from their urban values.

Legibility

The designed form of the public space has to meet one other indispensable condition: to be easily readable, to be comprehensible. If this is not so, if the citizens do not have the sense of being carried along by spaces which communicate their identity and enable them to predict itineraries and convergences, the city loses a considerable part of its capacity in terms of information and accessibility. In other words, it ceases to be a stimulus to collective life.

To establish a comprehensible language it is necessary to reuse the semantics and the syntax that the citizen has already assimilated by means of the accumulation and superimposition of the terms of a traditional grammar. It is not a matter of simply reproducing the historical morphologies but of reinterpreting what is legible and anthropologically embodied in the street, the square, the garden, the monument, the city block, etc.

No doubt with these ideas I will be accused by many supposedly innovative urbanists of being conservative, reactionary, antiquated. But I want to insist on the fact that the city has a language of its own which it is very difficult to escape. It is not a matter of reproducing Haussmann's boulevards, or the street grids of the nineteenth century, or Baroque squares or the gardens of Le Nôtre. It is a matter of analysing, for example, what constitutes the centripetal values of these squares, what is the plurifunctional power of a street lined with shops, what are the dimensions that have permitted the establishment of the most frequent typologies. And it is a matter of being aware of how abandoning these canons results in the death of the city [and leads to] the residual spaces of the periphery and the suburb, the vast shopping centres on the outskirts of the city, the urban expressways, the university campus at a considerable distance from the urban core, and so on.

Architectural projects versus general plans

The above considerations bring us to another important conclusion which has been applied in Barcelona: the urbanistic instruments for the reconstruction and the extension of a city cannot be limited to normative and quantitative general plans (Figure 5.2). It is necessary to go further and give concrete definition to the urban forms. In other words, instead of utilizing the general plans as the only document, a series of one-off urban projects have to be imposed. It is a matter of replacing urbanism with architecture. It is necessary to design the public space — that is, the city — point to point, area by area, in architectural terms. The general plan may serve very well as a scheme of intentions but it will not be effective until it is the sum of these projects, plus the study of the large-scale general systems of the wider territory, plus the political definition of objectives and methods. All over Europe during these past 30 years general plans have justified the dissolution of the city, its lack of physical and social continuity, its fragmentation into ghettos, and have paved the way for criminal

5.2
Olympic Port

speculation in non-development land. And they have, in addition, counterfeited a spirit of popular participation, whose criteria cannot logically be extended beyond the local neighbourhood dimension.

The continuity of the centralities

Controlling the city on the basis of a series of projects rather than uniform general plans makes it possible to give continuity to the urban character, the continuity of relative centralities. This is one way of overcoming the acute social differences between historic centre and periphery.

I am aware that in these past few years many voices have spoken out in defence of the diffuse, informal city of the peripheries as the desirable and foreseeable future of the modern city. The *ville eclatée*, the *terrain vague*. This position seems to me to be extremely mistaken.

The peripheries have not been built to satisfy the wishes of the users. They have appeared for two reasons: to serve the interests of the capital invested in public or private development and to further conservative policy. They exploit the value of plots outside of the area scheduled for development, and they segregate from the main body of the community those social groups and activities regarded as problematic by the dominant classes.

Urbanists who uphold the model of the periphery seem not to realize that all they are doing is putting themselves on the side of the market speculators, without adding any kind of ethical consideration. As certain neo-liberal politicians say, the market takes over from policy: without considering the economic and social damage suffered by the periphery and even by the suburb — in other words, without culture, without politics.

Architectural quality between service and revolutionary prophecy

No urbanistic proposal will make any kind of sense if it does not rest on architectural quality. This is a difficult issue. If the city and architecture are to be at the service of society, they need to be accepted and understood by society. But if architecture is an art, a cultural effort, it must be an act of innovation towards the future, in opposition to established customs. Good architecture cannot avoid being a prophecy, in conflict with actuality. On the one hand actual service in the here and now, and on the other hand anti-establishment prophecy: this is the difficult dilemma which good architecture has to resolve.

Architecture as a project for the city

I do not want to conclude without referring to another architectural problem. It is evident that these days there is a great split in the diversity of architectural output. On the one hand there is the tightly rationed production of the great architects which is published in the magazines and shown in the exhibitions. On the other hand there is the superabundance of real architecture, that which is constructed in our horrible suburbs, along our holiday coasts, on the edges of our motorways, in our shopping centres. A very bad architecture, the worst in history, which destroys cities and landscapes.

There are many reasons for this phenomenon, but the most evident ones are the typological peculiarity of the great projects and the commercialization of vulgar architecture. The great Ivory Tower projects are no longer capable of putting forward methodological and stylistic models, and as a result the majority of vulgar architecture cannot even resort to the mannered copy.

Clearly, we are not in any condition today to call for the creation of academic models, as has occurred in the history of all styles. Perhaps the only possibility open to us is that of establishing a rule that is more methodological than stylistic: that architecture should be primarily a consequence of the form of the city and of the landscape and should participate in the new configuration of these. This would be a good instrument for a new order, in opposition to the self-satisfied lucubrations of good architecture and the lack of culture of vulgar architecture.

I began by saying that the city must be an architectural project and I have ended by saying that the solution to the present problems of architecture may be to design it as part of the city.

Note

This chapter was published in *Architectural Review* in September 1999. We thank the publishers for permission to reproduce it here.

Chapter 6

The city, democracy and governability: the case of Barcelona

Jordi Borja

Introduction

Three urban and political processes come together in Barcelona: the democratization of municipal government, the physical transformation of the city, and the formulation of a city project that meets with the broad approval of civil society. In this chapter I shall analyse these processes from three standpoints: first, local government initiatives: second, the links between public authorities and the private sector: and third, the challenges to future governability.

Municipal government initiatives

The 1979 elections, whose results were later confirmed in 1983, 1987 and 1991, gave a clear-cut majority to the forces of the left, which could also count initially on the support of the centre. The right-wing opposition (*Partido Popular*) has always held only a small minority of seats on the city council. The governing majority has always benefited from the strong personal leadership of the mayor.

A municipal government was set up with a left-wing majority, but including the centre (the "pact for progress", representing over 90% of the vote). Although the centre later left the governing majority, major municipal policy issues have always been settled by agreement (the centre has governed the region since 1980).

Administrative reform, which was introduced immediately, cut back and professionalized local government staff, streamlined administration, computerized services and encouraged functional decentralization through the establishment of independent enterprises and bodies. On this basis the city administration was able to enhance its credibility and efficiency, initially to impose standards and discipline on the private sector (for example, contractors and licensees) and at a later stage to promote cooperation.

Geographical decentralization through the creation of districts was undoubtedly the most important political act of the early years of municipal democracy, together with the urban policy of creating public spaces and amenities in all neighbourhoods. Decentralization had three consequences: it opened the way to integrated neighbourhood activities, brought the administration closer to the people and their requirements, and made municipal policies more responsive to the needs of peripheral and working-class areas. Political and administrative decentralization paved the way for the development of citizen participation, facilitated relations with community bodies, and greatly increased opportunities for contact with the citizens.

Development of public amenities and promotion of the city

The democratic city government quickly made a special effort to restore or create an attractive urban environment by many different means: improving the city's appearance and security, creating small parks wherever possible, encouraging neighbourhood and city-wide festivals, reviving traditions (such as processions and carnivals) and other similar activities (Figure 6.1). The city's central and local parks were put to a variety of uses. At the same time, internal

6.1
Civic (community) Centre in Carmel neighbourhood

6.2
New waterfront promenade next to Barceloneta, built in the mid-1990s

promotional campaigns were launched, some of a general kind ("Barcelona for ever"), others having to do with urban renewal ("Barcelona, look your best") and yet others involving public services (markets, public transport, clean streets, and so on). When the campaigns were of a general kind, they were always linked to actual events, such as Barcelona's bid for the Olympic Games, or, in the case of more specific campaigns, to improved services. Cultural projects (such as museums, exhibitions and concerts) and sporting events — later to culminate in the 1992 Olympic Games — constituted a particularly important feature of this policy of integration and promotion.

The city's social policy included a "hardware" element involving the creation of public spaces and amenities in all neighbourhoods and the decentralization of cultural policy and social services (Figure 6.2). A programme of urban investment in the neighbourhoods was carried out during the first year of democratic rule. Decentralization began with social services (including administrative and information offices). This policy, which was an immediate success and enjoyed widespread public and even international recognition (the neighbourhood amenities programme won the Prince of Wales' prize for town planning), played a decisive part in the subsequent development of the city. It gave rise to a considerable improvement in the quality of life of the population and in the functioning of the city and its services, while demonstrating that in spite of scarce resources it was possible to achieve a great deal in the most run-down areas. The principal economic constraint was not investment in construction but maintenance costs. The suburbs attained city status, and great care was taken over both the functional and the aesthetic aspects of the projects. The decentralization of social and cultural services combined with town-planning activities in the neighbourhoods helped to promote both their internal integration and their civic solidarity. The foundations of a strong social consensus were laid. But the policy also contributed indirectly to an economic revival by enhancing the urban environment, making it more suited to new tertiary activities, and upgrading human resources by giving a sense of citizenship to the inhabitants of working-class areas (Figure 6.3).

Not for lack of goodwill or good intentions, the attempt to achieve greater metropolitan cohesion was not the city government's most successful venture. From the outset Barcelona council relinquished the position of legal pre-eminence it held in the metropolitan corporation, which linked the city to the 26 municipalities of the inner ring, in favour of an assembly of mayors. With the support of the left-wing majorities in power in all the municipalities, a policy of metropolitan coordination was introduced to promote "city-building" activities in the metropolitan area, covering such fields as communications infrastructures, parks and amenities, improved basic services in respect of water, sanitation and refuse, development of the intermodal transport system, joint economic promotion activities such as the technology park, and so on.

6.3
New municipal market, Sagrada Familia neighbourhood, built late 1990s

This policy gained momentum when the major projects for 1992 were being drawn up. The ring roads building exercise, for example, was planned as a major urban development scheme for the inner ring. However, the political stand-off with the regional government of Catalonia, which was concerned that a "dual authority" might result from a strong metropolitan government, and the scant enthusiasm of the peripheral local authorities for strengthening a metropolitan structure that would inevitably be dominated by Barcelona council not only prevented any expansion of the metropolitan structure, establishing it as a metropolitan region (i.e. including the outer ring), but led to the break-up of the existing body. A controversial law passed by the Catalan government in 1987 dissolved the metropolitan corporation, established a series of supramunicipal bodies and took over the decisive town-planning powers (such as the approval of plans). Since then a de facto metropolitan policy has continued to function but without the requisite institutional and legal framework.

Urban aesthetics

The aesthetics aspect of urban projects, the care given to the design of the city's features and image and the attention paid to all artistic and cultural manifestations are not just a question of marketing (although marketing comes into it), much less one of following fashion or giving the professional a free hand. Urban aesthetics fulfil three functions, the first being to give the city, as a whole and at neighbourhood level, a sense of unity. Monuments and sculptures (through what they represent and the reputation of their creators), the formal beauty and originality of design of city infrastructures and facilities (for example, certain bridges, renovated stations, the sports stadium), and the grooming of city squares and gardens, give greater dignity to the city's inhabitants, raise its profile and strengthen the identity and civic pride of the population. Urban aesthetics help to shape the cultural references that citizens must have if they are to feel at home in their city.

The second function is to demonstrate the quality of public administration. The aim is to produce work and services that are not just routine, but of the highest possible standard. Precisely because it is not a conventional obligation, aesthetic concern is indicative of the administration's commitment to work well done and helps to make the city more egalitarian and community-minded. This concern for quality, regarded by the administration as an obligation, generates a sense of duty and civic responsibility among the population. The design and presentation of a metropolis, which may seem secondary, can prove to be of the utmost importance in giving meaning and visibility to the city as a whole.

The third function of urban aesthetics is to bring out the most distinctive and attractive features of the city — an invaluable element in city marketing. Nowadays the urban environment, cultural facilities and profile or image of a city are of vital importance to attract visitors, tourists, convention-organizers and investors. The embellishment of a city is not just a good thing: it is a sound investment.

Coordination amongst the parties concerned

Very little of what has been done would have been possible without the development of coordination mechanisms between administrative bodies and the economic and social forces. Certain aspects of this coordination deserve attention.

The activities of the decentralized geographical units (i.e. districts or large neighbourhoods) were carried out partly through cooperation with the population. The civic centres (socio-cultural amenity centres providing a wide variety of service and activities, established in all neighbourhoods) are run jointly by the district and community organizations and associations. Many of the services planned and financed by the municipality are managed indirectly through community bodies, for example the organization of festivals, literacy and Catalan courses, home help for old people and invalids, cultural activities, the minding of public precincts, and so on.

Urban renewal and development projects have provided many opportunities for cooperation between the public and private sectors. The "Barcelona, look your best" campaign is a major promotional venture on the part of the municipality, which does the advertising, seeks sponsors to renovate important buildings and gives economic assistance to individuals. The campaign has encouraged thousands of citizens to repaint buildings, overhaul services and modernize their homes and businesses (some 5,000 such projects have been, or are being, carried out).

A further type of activity, known as "coordinated town planning", enables ambitious operations to be undertaken along the lines of the "new centralities" and Port Vell (Old Port) projects, in which an initial public operation encourages the private sector to submit "project plans" that meet public requirements of quality and balanced use while permitting work to begin at once. The major post-1992 projects (such as the Diagonal Mar on the waterfront, created as a result of the Olympic Village and the ring road along the coast), depend largely on the private ventures triggered by the initial public operation. The limitations of such cooperation emerge later: the private sector will not move until the short-term profitability of the operation is guaranteed. If the economic climate does not seem very favourable, development is slower than expected.

The municipal government has mounted several campaigns calling for public cooperation in such areas as the popularization of sport and the fight against drug addiction. In other cases it has supported campaigns initiated by civil society, such as those aimed at combating racism and xenophobia. It is interesting to note the effective cooperation mobilized by international solidarity campaigns, like the one on behalf of Sarajevo, through which the municipality collected ten times its own contribution.

Another type of cooperation between the public and private sectors sprang up for the worldwide promotion of Barcelona. Some permanent joint institutions and societies were created, such as the Tourist Consortium set up by the municipal government, the hotel owners association, the Chamber of Commerce, the Trade Fair and similar bodies, and temporary associations, such as the European Agency for Medicines and the European Federal Bank, were set up for specific purposes.

As regards cooperation between the public and private sectors in business and job creation, the municipal government took innovative action more in response to the acute crisis of principles that arose in the 1980s than out of any ideologically motivated or doctrinaire interventionist policy. This action took various forms: the promotion of innovative companies (e.g. Iniciativas S. A.), training for entrepreneurs and assistance for the creation of small-scale enterprises (Barcelona activa), training for young people and advice for very small business projects (through local development agents in the neighbourhoods). These municipal government activities invariably gave rise to cooperation between the public and private sectors in financing (e.g. Iniciativas S. A. and Barcelona Impuls acting as a risk-capital bank) and in management.

The Olympic Games operation (or more accurately the major infrastructural projects carried out in connection with the games) gave further impetus to collaboration among the public administrative bodies, which is almost invariably more difficult to arrange than collaboration between the public and private sectors. Collaboration among administrative bodies was based on three simple, practical measures: a schedule of the activities to be carried out, a breakdown of funding among the various bodies involved, and a system of public enterprises to implement the activities (in the form of a financial holding company and a series of executing companies). Cooperation between the public and private sectors was then developed on this basis to provide investment opportunities involving the return or repossession of the facilities, such as the Olympic Village, telecommunications towers and service galleries, once the games were over.

Cooperation between the public and private sectors was also enlisted for the building and operation of such major economic infrastructures as the trade fair and the conference centre (still at the planning stage). The Municipal Charter (or special law of Barcelona) provides for civic (joint) management of the major economic and communications infrastructures (such as the airport and the harbour). An effort was made from the very outset to associate private groups with the financing and running of such large-scale cultural projects as the Liceo or opera house and the museum of modern art.

The strategic plan is undoubtedly the most accomplished product of cooperation between the public and private sectors. In the context of the major projects of the late 1980s and the run-up to the 1992 Olympic Games, an urban structure was devised that called on the principal public and private partners in the business community, the trade unions, the professions, and cultural and social activities, to analyse the situation, formulate joint objectives and propose strategies and action for the period from 1992. The result was the Barcelona 2000 Strategic Plan. The first version, approved in 1991, was followed by a second in 1995, indicating that this is a lasting undertaking and not just an offshoot of the enthusiasm that presided over its birth. The Strategic Plan is not a legally binding instrument but a "political and social contract" whose execution is the responsibility of those who are competent and qualified to assume it. At the same time, it offers a means of applying public pressure to secure the fulfilment of its objectives.

The challenges to future governability

The city has shown itself to be highly governable, with a relatively low level of social conflict (in spite of unemployment, crisis in the automobile sector and the weight of a bureaucratized public sector) and a high degree of cultural integration. It is easy to be reassuring, even overconfident, about urban society and local government in Barcelona. However, I believe it would be more constructive to emphasize the political and social problems that will require firm and innovative responses in the next few years since they affect the very viability of the city as a social and political organization. Such responses are possible, although by no means easy.

The present system of political organization and distribution of responsibilities does not guarantee governability, representativeness or efficiency. As the political leader, the mayor should be able to count on a clear-cut majority in the city council. This has been available up to now, but the present proportional system does not guarantee it. On the contrary, it tends to depersonalize political representation.

Decentralization also depends on the direct election in the districts of both the chairperson, who should be the councillor receiving the most votes, and the councillors, as laid down in the decentralization regulations approved in 1986, but not yet applied in this respect. Electoral competition at city level is unlike that between national political parties. It would be better if political parties did not compete as such in municipal elections. Local government should be able to assume new responsibilities in matters relating to economic affairs, the management of major municipal services, justice, security, and so on. The coordinating role of local government in relation to other public bodies is not officially recognized, although it has been of great practical importance. And there is the whole field of contractual relations between administrative bodies (in addition to cooperation between the public and private sectors) which requires closer regulation based on the principles of proximity and integrated action.

Social disruption is a real threat arising out of current forces of exclusion, although they have been countered in the present decade by more potent integrating forces. Three of the forces making for exclusion, namely unemployment, housing and the marginalization of young people, are particularly powerful and dangerous, mainly because they are mutually reinforcing. Unemployment demands a "municipal" response, since we know that market forces are unable to cope. Despite the dynamism of the past ten years, Barcelona, with an unemployment rate of over 10% in the metropolitan area, will have to "invent" jobs in such areas as urban ecology, the maintenance of infrastructures and services and in the so-called "proximity" sector (jobs of a social and cultural nature, involving personal and other services).

The supply of housing for young couples is highly inadequate, as are efforts to make the housing market, both primary and secondary, more flexible. It has to be acknowledged that public housing policy — a responsibility of regional government — has been exceptionally weak in recent years.

Lastly, the marginalization of young people has up to now been balanced by a strong trend towards social and cultural integration within the city, as demonstrated, for example, by the tens of thousands of Olympic volunteers, the spectacular spread of sporting activities and, on a different plane, the vitality of the NGOs working for international cooperation and intercultural solidarity. Nevertheless, the growing alienation of the young from political institutions and their vulnerability to unemployment and lack of housing make it essential to keep a close watch on a problem that is likely to grow more acute.

The demand for civic security and justice is very strong, far exceeding the ability of public institutions and authorities to meet it. Experience in Barcelona has led to three conclusions. Local government must assume responsibility for these matters. Security calls for coordination at the local level and cooperation on the part of the populace. The municipal administration of justice can help to relieve the pressure on the present judicial system by resolving conflicts through conciliation and arbitration, punishing minor urban delinquency and intervening in the city's own problems, such as housing and consumption. There is a need to identify new criminal acts (such as damage to the environment, racism and xenophobia) and to punish them visibly and effectively, while at the same time decriminalizing others forms of behaviour, such as drug addiction.

Citizen participation is a key element in the building of secure and just cities. In some cases novel practices have emerged, such as civic security councils (at city-wide or district level), community policing, free distribution by bus of methadone, tolerance of private drug use and non-judicial punishment of certain kinds of antisocial behaviour, such as vandalization of public property, and air, water and noise pollution. In other cases, proposals have been drawn up and incorporated in the draft Municipal Charter for the appointment of district judges and prosecutors or the assignment of responsibility for coordinating policing in the city to the mayor.

At all events, we should not forget that an inadequate response to the demand for justice and security leads to the development of "tribal" behaviour and antisocial attitudes. In Barcelona this response is based on the principles of proximity and cooperation, combining preventive measures with suppression and re-education (with security councils, community policing and municipal justice).

Local identities, multiculturalism and universality

The globalization of the economy and the worldwide spread of the media lead to greater cultural homogeneity and a quest for local identity. At the same time, the combination of increasingly fragmented social structures based on differences in status with the influx of impoverished ethnic groups gives rise to defensive reactions on the part of integrated but vulnerable groups threatened by loss of identity and security as well as by unemployment and an uncertain future. In the case of Barcelona two factors have contributed to the process of integration. The first was the assertion of Catalan identity at the historic moment when political independence was being regained and Spain was undergoing democratization and integration into Europe. This assertion of identity, incidentally, was not aggressive, nor did it lead to defensive isolation.

The second factor was the upsurge of "civic patriotism" in Barcelona, linked moreover to the "internationalization" of the city. "Putting the city on the world map" was one of the municipal government's objectives, involving a major "city marketing" operation. Furthermore, the majority of citizens saw the Olympic Games as an opportunity for scoring a success and obtaining recognition on the world scene. And these two factors were combined with an array of public policies, as outlined above, designed to establish a constructive dialectic between neighbourhood identity and the overall concept of citizenship. This favourable situation, however, should not conceal the fact that contradictions do exist and that the present equilibrium is precarious. The erosion of the comprehensive city project, rising unemployment, the current increase in the number of racist acts and growing geographical and social inequalities — all of them attested in the post-1992 data — might well trigger social and cultural conflicts that have so far been negligible. Barcelona has not yet undergone any major local conflicts as a result of "exclusion" or "asymmetry", but this does not mean that we are immune to them.

Social disintegration and citizen participation

One of the greatest achievements of this period is undoubtedly the public acceptance of the city project. Barcelona cannot be said to have escaped the processes of disintegration at work in the other major cities of the world, such as a rapid increase in the number of people living alone, low membership of trade unions and political parties, or greater reliance on family than on voluntary community organizations. Nevertheless, the compounding of the democratic civic mobilization of the 1970s with the integrating capacity of the city project of the 1980s generated a number of countervailing forces. We shall consider three of the most significant of them.

- The first is the relative strength of associations in the neighbourhoods: the inhabitants of Barcelona have twice as many associations as the metropolitan area and four times as many as other major Spanish cities.
- Attention should also be drawn to the strength of the decentralization process, which recognizes political pluralism (the district chairperson represents the majority group in the area even if it is a minority in the city as a whole, which is the exception in Spain and not very frequent in Europe) and has given rise to systems of representation and provision of services that multiply contacts between the citizen and the administration.

- And lastly there is a functioning network of citizen participation and communication mechanisms, ranging from geographically and sectorally based participation councils to local radio and television stations and the on-going cabling of the city.

To all this should be added a large number of novel procedures to facilitate relations between the public and the administration (such as conducting business by telephone, single-window operations, and the proposal to give oral declarations the status of official documents). Nevertheless, we cannot afford to overlook the strength of the processes of disintegration or to ignore the fact that the processes of participation arose in response to political circumstances and convictions that are by no means certain to survive. It is therefore essential to firmly establish a legal and political framework embodying the innovative processes of the past decade.

The Municipal Charter: a political and legal framework

It is a truism that present-day political and legal systems are inadequate to the task of governing our major cities. Everything is subject to criticism: failure to recognize the multi-municipal nature of metropolitan cities, the mayor's lack of executive power or, on the contrary, the inability of the council to intervene, "unbusinesslike" administrative procedures that stipulate multiple prior controls but only perfunctory evaluation of the results, a narrow range of responsibilities resulting from gradual attrition (in housing or social services, for example) or because some areas are not covered by existing regulations (for example, justice, international promotional activities and job creation), the failure of other administrative bodies to make contracting compulsory or to coordinate the activities of other administrative bodies with those of local government, and the difficulty or impossibility experienced by municipal authorities in giving official status to novel procedures in such matters as cooperation with the private sector, town planning and the spread of new communication technologies.

The Barcelona city government has introduced a series of comprehensive and intensive measures that have enabled it, under favourable political and social circumstances, to take action in various unregulated areas and to make innovative proposals concerning every aspect of local administration. The tangible result of these activities is the draft Municipal Charter or Special Law of Barcelona, drawn up in 1990–91 and approved by the municipal council in 1994. This draft proposes specific regulations for the city, based on principles that could also apply to other large cities.

The Charter takes up issues of political organization, such as the introduction of a majority element into the election of the mayor and district chairpersons and the assignment of greater powers — including decision-making — to the committees of the city council. The decentralization process is further strengthened by the direct and individual election of district council members. All the various forms of participation and the greater flexibility given to relations between the public and private sectors that have been tried out in recent years are institutionalized. The new town planning and management instruments that are already in use, such as the Strategic Plan and project plans, have been made official, and local government is given the power to approve town plans, determine land use and requisition the city sites and run-down installations belonging to other administrative bodies (for example railway and harbour zones, other sites, public works areas, and so on).

Contracting is made compulsory for certain activities and services, such as access infrastructures and public transport, and local government is given the right to coordinate activities with other administrative bodies within its geographical area. New responsibilities (including justice, basic education, economic development and the administration of housing programmes) are handed over to local government, and its financial resources are stepped up in consequence.

The force of these proposals resides in their proven worth. In other words, they are not based on ideological or doctrinaire grounds but are the fruit of more than ten years' experience in carrying through a process of change, in the course of which various instruments and procedures have been tested, numerous innovations introduced and some activities set aside for lack of means. Failure to adopt the Special Law or Municipal Charter, or its gradual erosion — the draft approved in 1994 is less ambitious than the 1991 version — would not only undermine future city governability but deprive the political culture of a political and legal frame of reference that could be of use to other cities.[1]

The organization of the metropolitan city

Cities today — big cities — are almost always multi-municipal. There are very few exceptions — one of them being Rome, whose municipal area of over 1,000 km² includes all of the actual city. Typically, we find the three dimensions of the urban phenomenon noted in the case of Barcelona: a central city (more or less coterminous with the city boundary), an agglomeration or urban continuum (the central city plus the municipalities of the inner ring, which almost always form the administrative area for basic services, such as transport and water), and the new metropolitan city (with the characteristics of the city-region).

The historical, political and functional complexity of the metropolitan city makes it extremely difficult to establish a viable government that is representative of the whole and has wide-ranging functions. In the case of Barcelona, a metropolitan policy was devised and put into effect by a wide variety of partners; its consistency was such that it resulted in a comprehensive city project (Project 92 and subsequently the Barcelona 2000 Strategic Plan). But this policy was the product of exceptional historic circumstances and relied heavily on the central city. It would seem impossible to guarantee the future development of the city-region without establishing a political and administrative structure that lays the foundations for the coherent administration of the whole area.

Our experience has shown that a metropolitan structure of this kind has to avoid three pitfalls that can lead to its non-viability or dysfunction. The first is the incremental tendency, i.e. the proliferation of public bodies of a political nature, with overlapping responsibilities and limited resources, which further contribute to the disintegration of the administrative area. This was the effect of Catalan law, which replaced the former metropolitan corporation by "special" and "local" bodies that were supramunicipal but submetropolitan. A further tendency is to eliminate or override peripheral municipalities in order to set up a local metropolitan authority identified with or dominated by the central city. This solution is politically unviable as a result of local and regional government resistance. A certain style associated with the initial years of democracy might have demonstrated this. Finally, there is a centralizing tendency, whose most striking expressions have been the abolition of the Greater London Council and the break-up of Santiago de Chile's metropolitan body during the dictatorship, in the course of which such metropolitan functions as area planning, water and sanitation, transport and refuse collection were transferred to the ministries concerned or to a central government agency.

Regardless of whether independent authorities or public-sector or mixed enterprises are set up to deal with each and every metropolitan service (this kind of arrangement, which includes contracting out to private companies, has proved expedient), experience shows that neither this nor any of the other methods referred to above can guarantee the systematic and democratic development of a metropolitan city. In our opinion, the best solution is to set up a consortium of public bodies, bringing together in a balanced way the regional government and representatives of the state ministries and independent authorities active in the metropolitan area with the mayors and representatives of the municipalities. This consortium would have three principal functions. First, it would draw up and approve a strategic plan and the sectoral and area plans relating thereto. Second, it would coordinate annual and longer-term investments by members of the consortium whenever they were of metropolitan or supramunicipal scale or impact. Third, it would be responsible for the political control of companies or independent authorities providing metropolitan services. Whatever the circumstances, no one can hope to monopolize regional metropolitan government. But it cannot be dispensed with, for not only does it supply the links between public authorities and serve as a focal point for private groups and individuals; it also ensures the visibility and hence the general accountability of public action.

Note

1   The Municipal Charter was given under a special law passed during the Franco years (in 1960) giving certain powers to Barcelona which other municipalities did not have (a similar law existed for Madrid). The proposed revision described in this chapter was approved by all political groups in the council at the end of the 1990s. But neither the Generalitat nor the central government have approved or ratified the text since then, and no new law has been passed.

Bibliography

Ayuntamiento de Barcelona (1990) *Plan Estratégico Económico y Social Barcelona 2000*, Barcelona.

Ayuntamiento de Barcelona (1990), *Barcelona y el sistema urbano europeo*, Barcelona: 'Barcelona Eurociudad', Collection, No. 1.

Ayuntamiento de Barcelona (1991) *Régimen especial para las grandes ciudades. La Carta Municipal de Barcelona*, Barcelona: 'Barcelona Eurociudad', Collection, No. 5.

Ayuntamiento de Barcelona (1994) Gabinete de Relaciones Exteriores, 1994. *Barcelona al món/en el mundo/in the world*, Barcelona.

Borja, J., Castells, M., Dorado, R. and Quintana, I. (eds) (1990) *Las grandes ciudades de la década de los noventa*, Madrid: Ed. Sistema.

Borja, J. (1987) Estado y Ciudad. PPU-Barcelona 1988, *Descentralización y participation ciudadana*, Madrid: IEAL-INAP.

Busquets, J. (1993) *Barcelona*, Barcelona: Mapfre.

Mancomunitat de Municipis de L'Area Metropolitana de Barcelona (1990) *Institut d'Estudis Metropolitans de Barcelona: Enquesta de la Regió Metropolitana de Barcelona 1990. Condicions de vida i hàbits de la població*, Barcelona (7 volumes).

Ministerio de Obras Públicas y Transportes (1992) *Estudios Territoriales*, No. 39 (Número monográfico: Una política para las ciudades), Madrid, May–August.

Programa de Gestión Urbana (UNDP-World Bank) (1995) *Barcelona, un modelo de transformación* Urbana, Quito.

Chapter 7

The planning project: bringing value to the periphery, recovering the centre

Juli Esteban

Introduction

The subject of this chapter is the planning project which was, and which continues to be, behind the urban transformation of Barcelona, something which is noticeable to everybody.

It certainly is not a project which is laid out in one specific document, nor which was conceived at the beginning of the process. Neither is it a project which has one, or several, exclusive authors, even though there are no doubt people who could be recognized as having played a particularly significant role in drawing up the project.

We could say that the planning project for Barcelona during the democratic period has taken form from a series of ideas and partial projects, with the participation of various actors and with different contextual references and specific goals throughout the period. This chapter will deal with the various initiatives and documents which make up this sequence of projects.

In particular it will deal with those projects dealing with broader planning and which therefore contain reflections of a certain breadth on the space which the city occupies. However, it should not be overlooked that the projects and instruments which are most immediately directed at private or public urban behaviour (projects involving construction, remodelling, renovation or maintenance work) also contain reflections which contribute to the overall planning project of the city.

In the development of the planning project, and naturally in the actual transformation which has taken place, there have been certain moments of special importance which have served to mark the pace of change. 1976, the year in which the General Metropolitan Plan (PGM) was approved, should be considered as the starting point of the process. 1979 was the year when the first democratic city council entered office. If it is important to note that in 1980 there was a key change in the city council urban planning team, with the substitution of Joan Anton Solans by Oriol Bohigas as director of planning and also the incorporation of Josep A. Acebillo into the team, the date which stands out as a reference point is 1979, marking the beginning of the process of democratic municipal management. The Olympic nomination received in 1986 is another key date in the sequence, although it should be understood that in terms of the project, the Olympic Games had been an objective on the horizon for several years — in 1982 the first draft of the Olympic proposal was drawn up, the Cuyàs report. Finally, 1992 is a date which does not need any further explanations and, as this date was passed, significant changes were introduced into the development of the project.

As is clear from what has been said up to now, the first instrument which we should refer to is the General Metropolitan Plan (PGM), which is the starting point, and a reference point maintained throughout the whole process, although its importance has changed as a result of actions becoming more distanced from it in time.

In what follows, three sets of projects are considered which were developed from 1979, as the unfolding and remodelling of the city's urban order. First of all there are the *localized projects*, which began to take shape from the first moments of the democratic city council and which brought with them the idea of proximity, of the recognition of the different identities of the urban make-up. After this the attention of the project moved on to consider *municipality-wide projects* which led to the proposal for new key nodes for this urban reality. At the same time, the existence of a *metropolitan administration* with metropolitan responsibilities allowed metropolitan projects to be taken on board. These projects, taken as a whole, constitute a complex and highly enriching deployment of the urban order proposed by the PGM, through which a new synthesis was produced. Its interrelated aspects expressed the coherent, evolutionary overall project which constituted at all times the planning of the city.

However, we can say that 1992 marked a qualitative change in the process. By 1992 all the projects with the characteristics mentioned above had been formed and partially carried out, and the pending projects were going through the natural process of organization and remodelling.

The period after 1992 is still too close to fully capture its significance. What we can identify is the variation in the operative circumstances and we can perceive some changes in the way the project was focused. Without renouncing the goals of the previous period, the project developed different perspectives on two classic urban dimensions: housing and industry.

On another level we could find those projects which were brought about through concrete actions, often more fragmented, especially at first, but which have proved to be a manifestation of ideas of the project as a whole and which have also been the broadest expression of how to carry out planning in Barcelona, to the point where they have eclipsed the significance of the project level of the planning.

In the city project which is found behind the practical projects we can identify two different but complimentary streams: those whose subject is public space, which range from the small squares and streets to the general systems of the city and the region, where what we can call public works are developed, and those which are concerned with the development and building of plots of land and which provide the volume of the urban make-up, where the work of private operators is of prime importance.

Having made these points on the organization of the project, we should comment on its objectives.

The most ideological documents on this theme are, on the one hand, the General Metropolitan Plan itself, and on the other hand, the texts which Oriol Bohigas provided during his term as director of planning, and in particular the introductory text to the publication *Plans i projectes per a Barcelona 1981–1982* (Plans and Projects for Barcelona 1981–1982) and the book *Reconstrucció de Barcelona* (The Reconstruction of Barcelona). Bohigas's texts represent two polarized and complementary positions; the first, from a planning perspective which lacked confidence in the projects which the PGM was to develop, and the second from a perspective of prioritizing action which is possible, necessary and which does not accept the overriding need for a general planning framework.

The plan for Barcelona has been built on this dialectic, through a process of successive dynamic equilibriums. If this has been possible, it is no doubt due to the shared objectives which the two ideologies have.

To bring value to, or monumentalize, the periphery and recover the centre is, without doubt, one of the expressions that best sums up the range of objectives which there have been in the plan for Barcelona.

We could argue that there exist basically two types of pathology in the city: on the one hand, the situation of degradation, in the sense of the loss of quality that the fabric of the city has suffered in certain central areas, and on the other hand the situation of the periphery which, due to many deficiencies, has not managed to become fully part of the city. In the Barcelona of the end of the 1970s, both pathologies were definitely intense and abundant.

The projects developed by the democratic city councils, with more fragmented views and sometimes in dialectical relation to the general planning schemes, were nonetheless fully coherent with the goal of recuperating the centre and monumentalizing the periphery.

If the most obvious and noticeable projects in this sense were to come from the action projects, and particularly from public works (squares, parks and the remodelling of various spaces), the projects that involved wider urban structuring have also been consistent with this goal.

Within localized planning, in the PERIs (Plans Especials de Reforma Interior, described later) we find alternately one or other intention, depending on whether they deal with the Raval inner city area or Vallbona on the periphery, for example. In the municipality-wide projects, the key objective is to distribute the potential for improvements in the areas which are qualitatively peripheral. Inevitably, the metropolitan projects are principally orientated towards ordering the topological, as well as the qualitative, periphery, as a large part of the coast and the Collserola range of hills had been until then.

The remodelling plans after 1992 are also part of this objective. It is no coincidence that two districts which may be considered as peripheral, Sant Andreu and Sant Martí, are the main focus of projects in this period.

In fact, the whole logic from which the actions of the public powers towards the city are derived, coincides with the goal of bringing value to the periphery and recovering the centre. The economic logic is to secure the correct efficiency of the existing urban "factory". The social logic aims to avoid the formation of enclaves and social segregation and contribute, both in the inner city and in the periphery, to the cohesion of the population. Finally, there is a logic of sustainability through which the full recovery of the city avoids the temptation of peripheral expansion, which consumes large amounts of energy and land.

As is indicated in the references, the literature on urban projects and actions which Barcelona city council and the metropolitan administration published on a regular basis, make available interesting information about many of the projects which are mentioned in the text.

The general framework: the 1976 General Metropolitan Plan (PGM) as a valid instrument for urban change

The city's project since 1979 has been based, to a large extent, on the historical circumstances surrounding the General Metropolitan Plan (PGM). This plan, taking in Barcelona and 26 surrounding municipalities, was finally passed in July 1976 and is still in force today.

The plan derives from planning law, that is, it was designed in accordance with the regulations laid down by legislation on land use and town planning. It must be remembered that, when the plan was being drawn up, the first Spanish land law was still in force, that of 1956, which was revised and replaced by a new structure in 1975, to which the plan's final version had to adapt.

The scope of the PGM has its origin in the 1953 plan, which was the first proposal that attempted to plan the city rationally in a space that significantly exceeded the limits of Barcelona municipality. However, this extension of limits was not the result of any rigorous study of relationships between different areas, but rather a last moment improvised proposal, due to a visit to the plan's headquarters by the director general of planning of central government. At the time, all decisions on planning were ultimately the responsibility of this director.

In spite of this origin, these boundaries have lasted until today through the PGM's prolonged existence. In addition, between 1974 and 1988, a metropolitan administration with planning powers existed; the Corporació Metropolitana de Barcelona (Barcelona Metropolitan Corporation, CMB). The metropolitan level actions have shown the possibility of the different areas working together; by having sufficient space they have improved the understanding of urban problems, although the choice of metropolitan boundaries is by no means optimal, not having been defined with sufficient scientific rigour.

It should be remembered that the size of metropolitan jurisdiction was seriously considered in the process of reflection on urban matters shared by the revision of the 1953 plan, and this was to be taken on board from 1964. This revision, leading to various progressive experts joining the planning process from outside the government bodies, gave rise in 1966 to the document entitled *Pla director de l'Àrea Metropolitana de Barcelona* (Regional Plan of the Metropolitan Area of Barcelona). This document contained a proposal for the territorial ordering of a much wider area, taking in the counties of Maresme, Vallès Oriental, Vallès Occidental, Barcelonès, Baix Llobregat, Alt Penedès and Garraf.

This document, given the name *Pla director* (Regional Plan) — a typology not considered in the urban legislation of the time, and which was then incorporated into the 1975 law on land use — also used for the first time the term *Metropolità*. Once the various administrations had got over their initial disquiet, a formula for the work to be done was agreed upon. The relevant administrative body, the Comissió d'Urbanisme i Serveis Comuns de Barcelona i altres Municipis, included Barcelona and the other 26 municipalities, the Provincial Council and central government. It should not be forgotten that the initial objective was to draw up a new legally binding plan for this area — with all that implied in terms of its detailed coverage.

In fact, the objective, explicitly stated in the documents of the Regional Plan, was to produce a General Plan under Spanish planning law for the entire area (the 7 counties, 164 municipalities). However, it was clear that a task of such scope could not be undertaken in one stage, and it was agreed to divide the development of the plan into the following sub-plans; the revision of the 1953 plan, covering the 27 municipalities; the plan termed "*immediate action*" made up of the rest of the counties of Maresme, Vallès Oriental and Occidental, and Baix Llobregat; and the "*deferred action*" plan made up of the counties of Alt Penedès and Garraf, to be developed at a later date. The elaboration of an infrastructure plan was also proposed, which would take in the whole area considered.

From the work on these fronts, several interesting products appeared, although the only one that was to reach its objective of setting up a binding urban plan was the PGM, which was initially approved in 1974 and finally passed in 1976.

It should be added that in 1974, shortly after the end of the new plan's public exhibition period, the Barcelona Metropolitan Corporation was created, considerably widening the powers of the Commission for Planning and Common Services (which had existed since 1953, the year of the previous plan's approval). This strengthened its role as the administration managing the metropolitan area. This is not to ignore its lack of democratic accountability, typical of the time, and its geographical restrictions, which were to last, because the 1953 limits were not reconsidered.

The role that the PGM was to have later in the democratic urban planning of Barcelona was partly due to its content, but also the result of the reading that was made of this plan. It should be remembered that certain events occurring between 1974 and 1975, along with the role of certain people, helped the establishment of this document as a basic reference point for urban management.

Although it was the beginning of the 1970s, the plan was not drawn up by officials supportive of the regime, but by those who at the time were called "infiltrated experts", with the help of professionals from outside the government (the same had happened with the Regional Plan). The result was consequently rather different from how official town planning had been up to then, as it included a strong corrective element in relation to the practices of town planning and building common at the time. It should be recognized that the powers-that-be did place their confidence in the experts and allowed them to work with a fair amount of independence. The decision to give initial approval to the plan in 1974, under the Barcelona mayorship of Enric Massó, was also important.

The process between 1974 and 1976, with two stages of public consultation, was especially interesting and was essential for the future of the plan, in addition to representing a valuable reference point for the revisions of General Plans that occurred later in many Catalan municipalities. The PGM received strong criticism both from popular sectors and from land and property owners. As still occurs quite regularly today with any plan over a certain size that is presented publicly, only some professional sectors fully aware of planning problems saw the plan in a positive light.

After a few months, the situation began to become clear and the plan received the backing of progressive sectors, a fact that was demonstrated by the dismissal by Mayor Viola of the head of the planning team, Albert Serratosa. He was at the time director of the Metropolitan Corporation's town planning services. In spite of this dismissal, and without

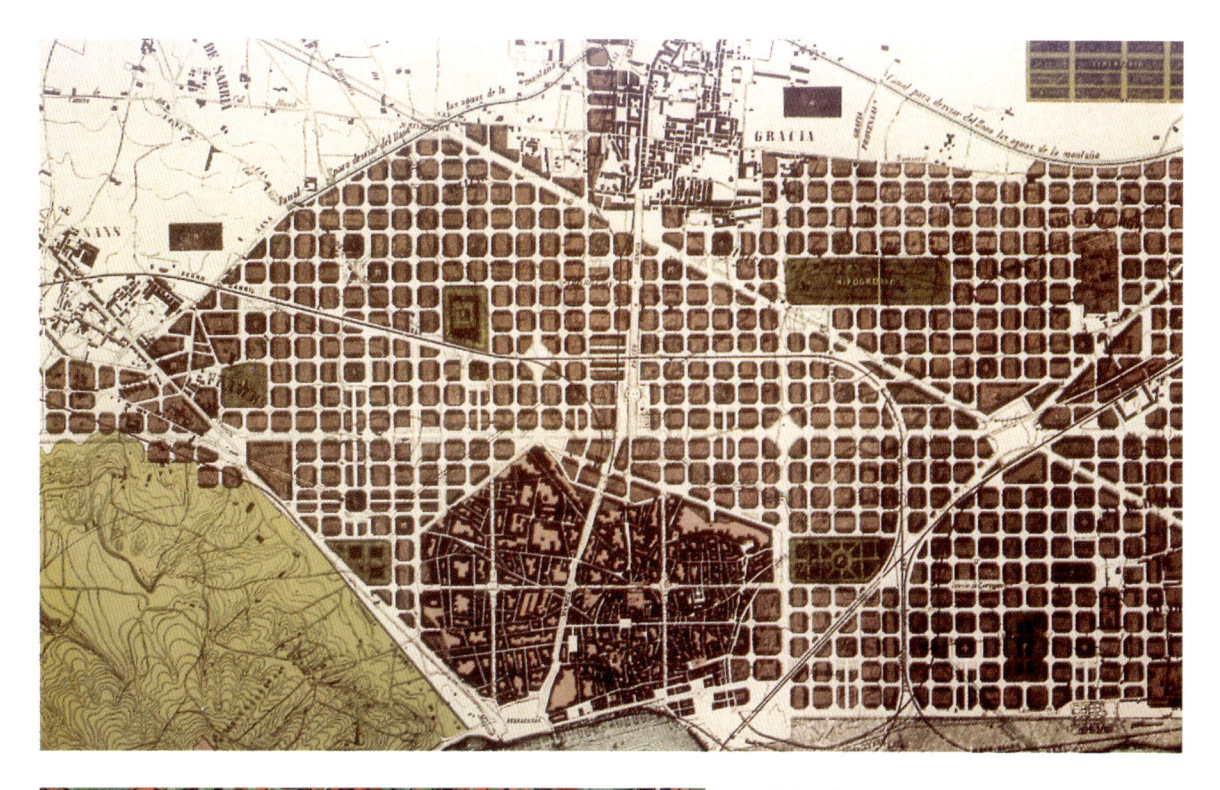

1 (*above*)
Cerdà Plan

2 (*left*)
General Metropolitan Plan (PGM) 1976 (central area)

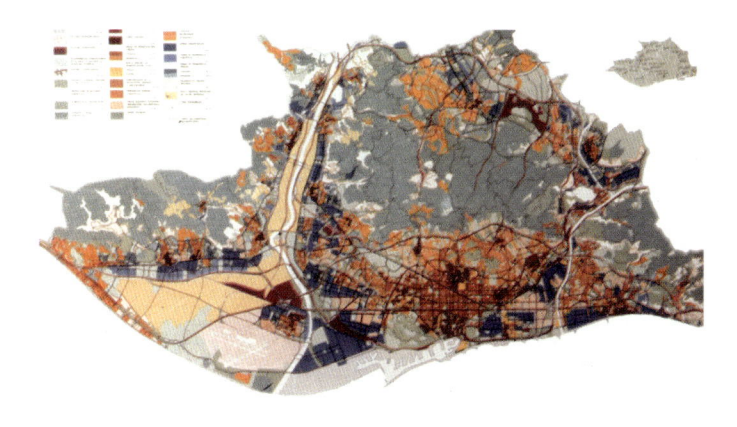

3 (*left*)
General Metropolitan Plan (PGM) 1976 (whole plan)

4 (*overleaf*)
Via Júlia after improvements in the 1980s

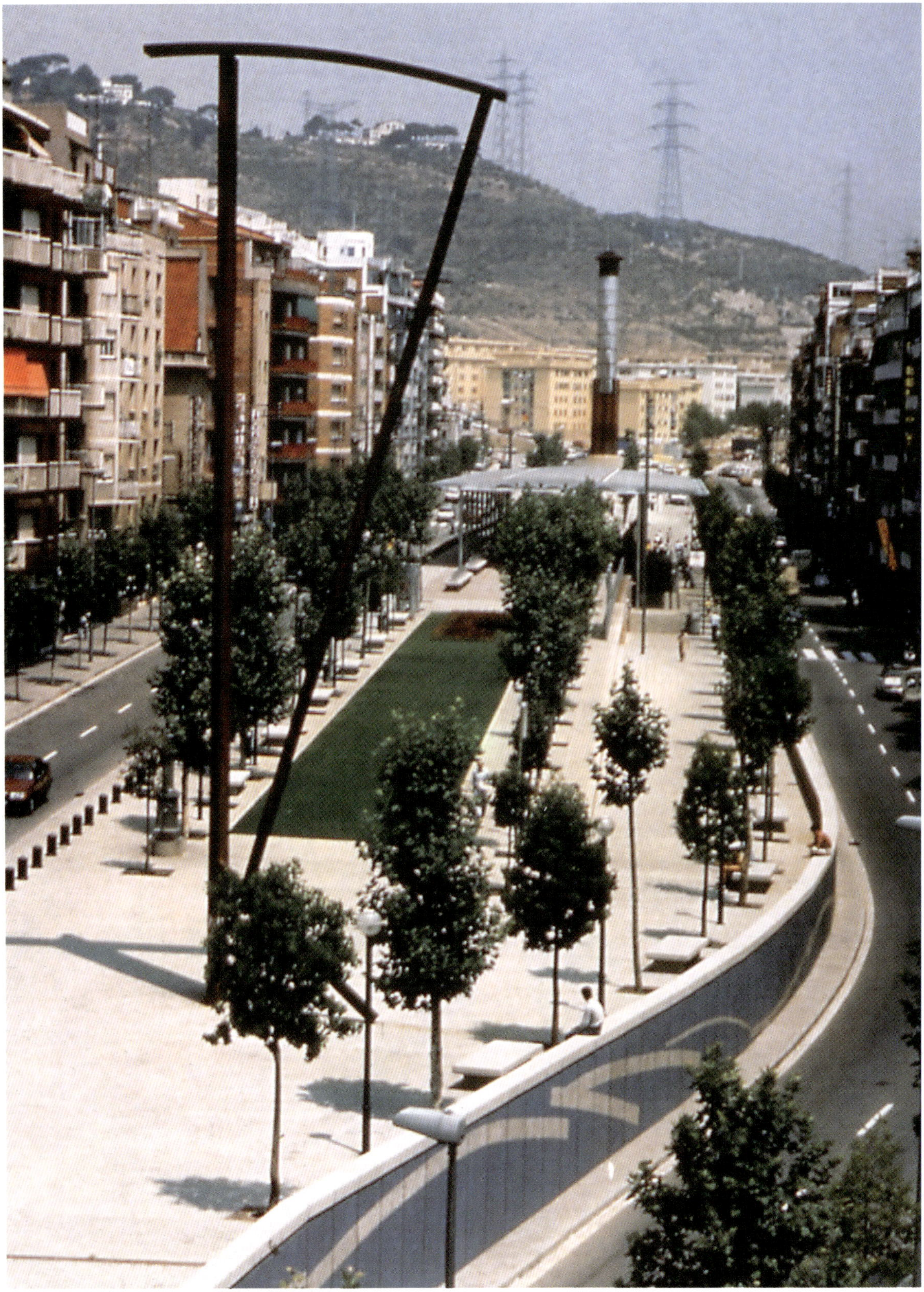

5
Open space on roof of building in Carmel neighbourhood

6 (*above*)
Maremagnum promenade, completed 1993

7 (*right*)
Rambla Prim, built in the 1980s

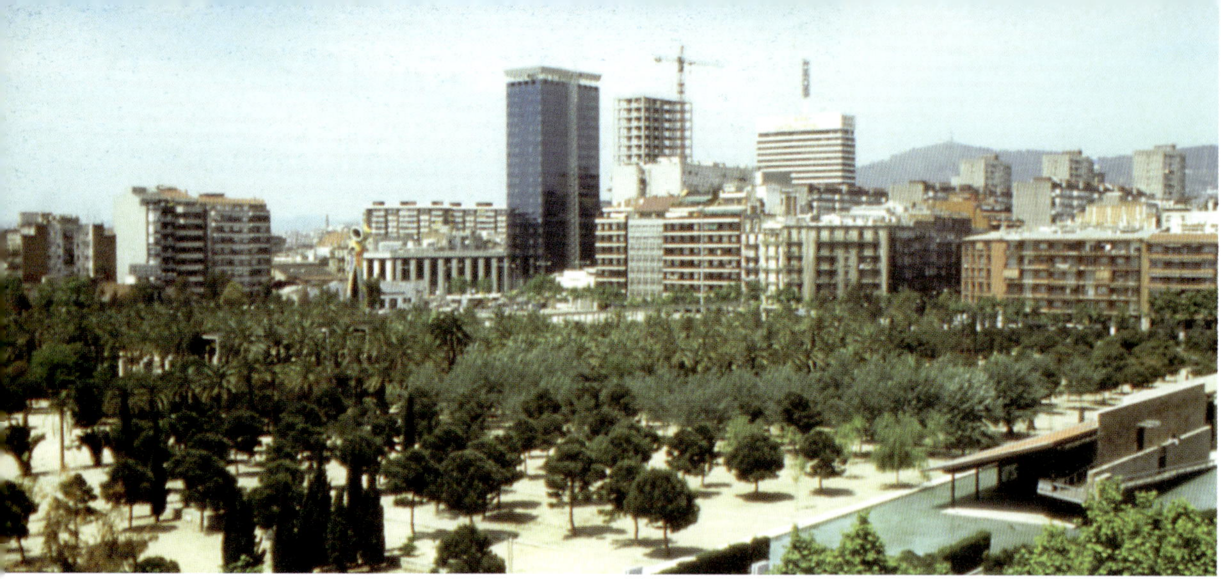

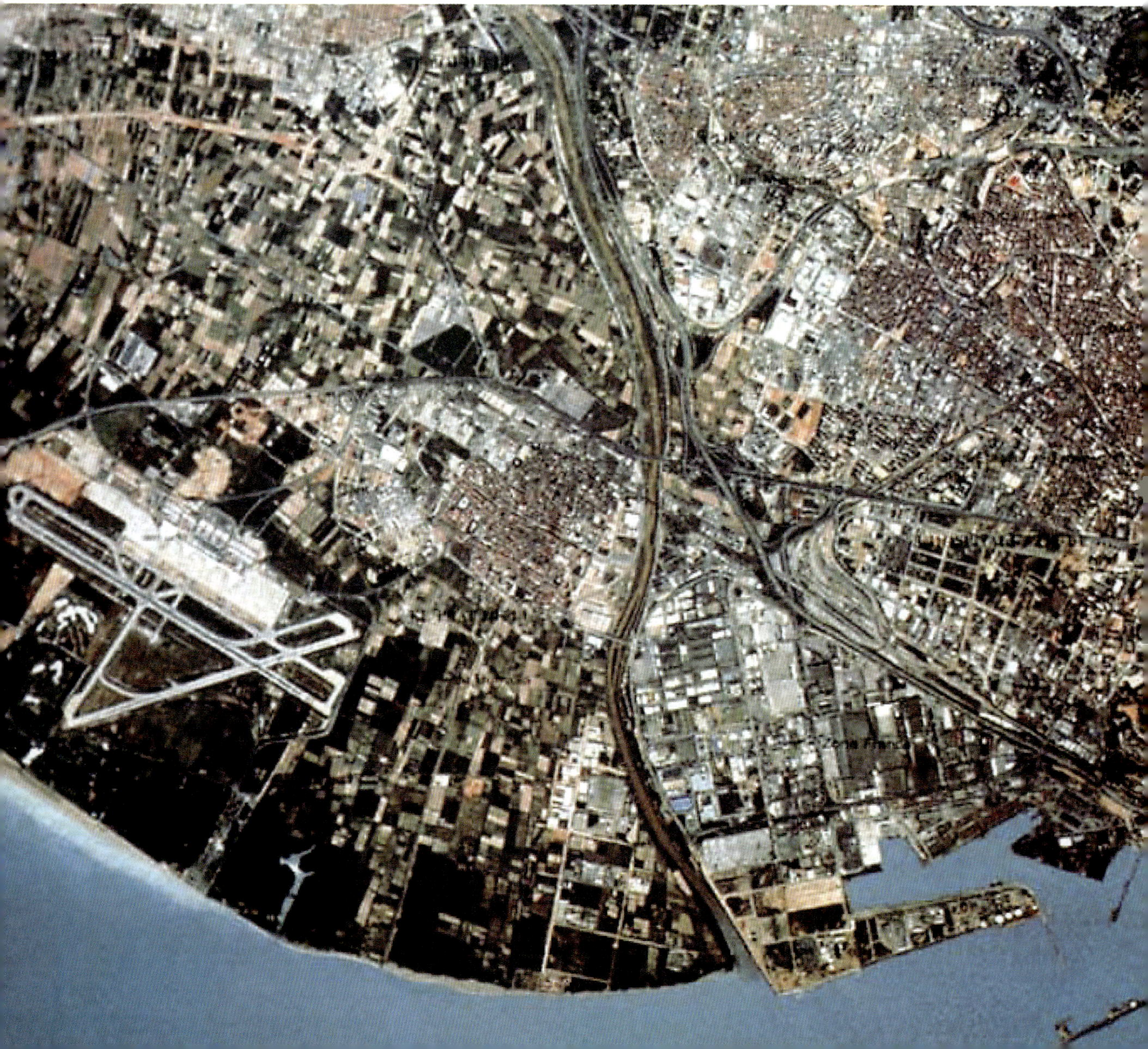

8 (*left*)
Joan Miró Park

9 (*right*)
Diagonal Mar Park

10 (*below*)
Aerial view of Barcelona coastline

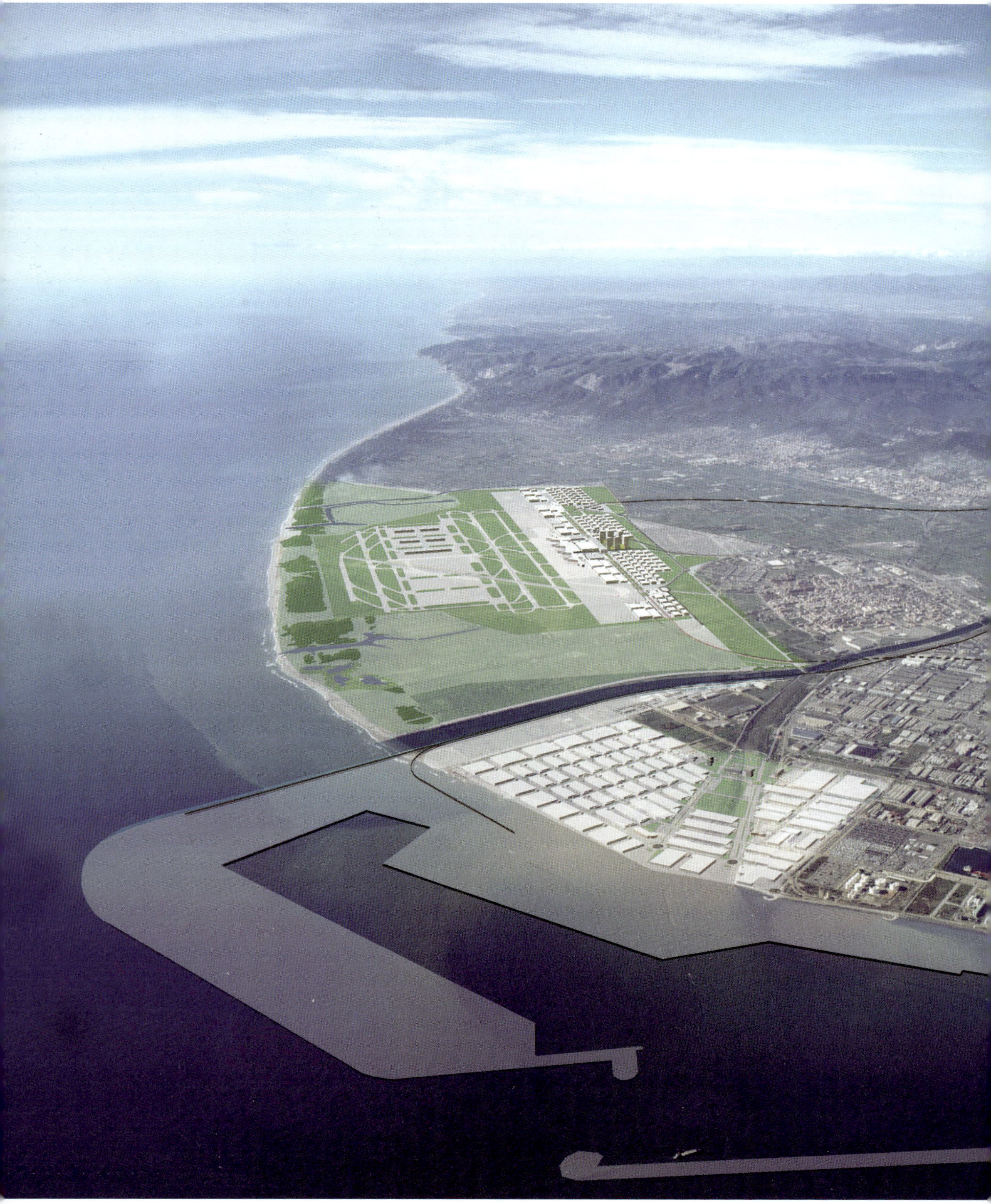

11 (*left*)
Aerial photomontage of proposed changes to port and airport, at mouth of river Llobregat

12 (*above*)
Aerial photomontage of schemes proposed at mouth of river Besòs

13 (*middle*)
Aerial photomontage of whole north western coastline, with schemes

14
Proposals for 2004 Forum, in earlier version

15
Plan of land uses in 22@ area (red is residential, blue is industrial)

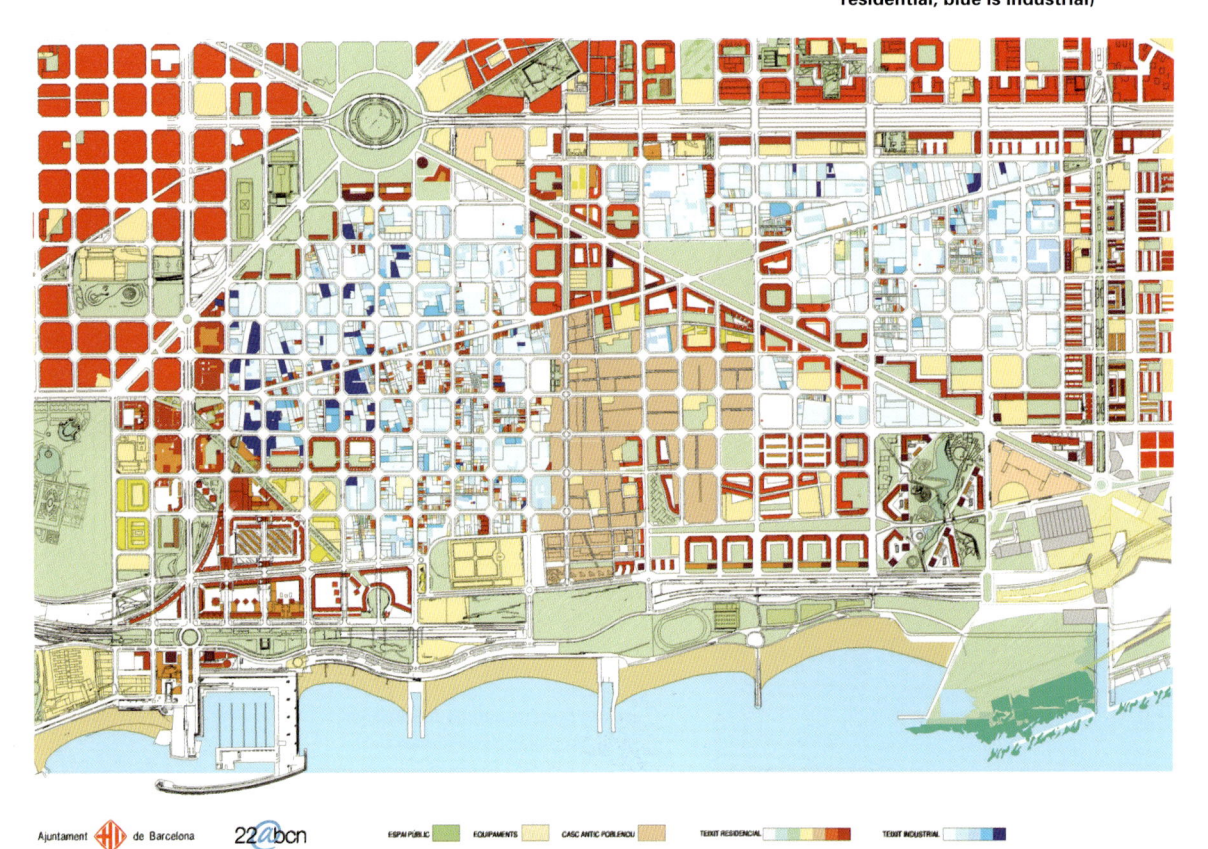

ignoring the various compromises and adjustments required by property interests which the plan had to introduce, it can be said that the plan that was passed in 1976 kept a good deal of the initial objectives and proposals. The reasons for this ability to resist the strong attacks on the plan undoubtedly lie in the level of maturity that the "social collective" had attained, allowing an implicit intelligence to deal with serious issues (and these were serious). This intelligence was demonstrated especially in the political transition that Spain underwent in those years.

In fact, this collective intelligence is what explains that the rest of the team drawing up the plan, led by Joan Antoni Solans, was not affected, and that the people who took over from Serratosa, Xavier Subias and Antoni Carceller, beyond their own sense of discipline, were competent experts with a clear understanding of the historic moment. Lastly, it was central government itself, through the official body based in Barcelona ("Órgano Desconcentrado") which had to give final approval to the plan and was presided over by the Civil Governor, Salvador Sánchez Terán, that amended some of the less acceptable concessions that had been made to property interests.

There are also some biographical circumstances that helped the acceptance of the General Metropolitan Plan by the first democratic city council. First, it was important that Joan Antoni Solans was named head of planning services (a position with significant room for action within this municipal organizational model) of the transitional city council presided over by Mayor Socias. Solans continued managing this body throughout the democratic city council's first year. In addition, it should also be noted that Serratosa formed part of the council as a local councillor, and that Pasqual Maragall, then first deputy mayor, had participated in the plan's last stage, specifically in drawing up the economic and financial plan.

At any rate, the most important thing is that the PGM, as a rupture with previous planning practices, was consistent overall with the attitudes of the new democratic councils.

However, due to the peculiarities of the plan's preparation, the PGM was, to a great extent, a technocratic proposal (in the good sense that the word had at that time) and in its approach expressed a clear mistrust towards political power. Specifically, it should be noted that the PGM imposed on the area a new legal planning framework, and to this end proposed a clear organization of public spaces or "systems". This was reflected especially in the road network, which was given an important place in the plan and affected numerous sites and quite a few buildings. As far as the impact of roads within urban areas was concerned, it should be said that the PGM maintained the same alignments as previous plans, which were consistent with the objective of reducing built-up density and improving the existing urban structure.

In addition, the PGM proposed a considerable quantity of land to be set aside for green zones and public facilities, which entailed the removal of numerous pieces of land from the property market.

As far as land for private initiatives was concerned (the "zones"), the plan worked towards containment. This was the case, first, in relation to the spread of urban sprawl into still rural areas and, second, through the substantial reduction in permitted building heights in urban areas, to avoid problems of building congestion.

It should also be added that the legal plan documents (on map bases and as texts) were significantly more precise than General Plans usually were. The plan expressed firmly a certain approach to the setting of the city on its territory and provided the regulations to carry this out.

Without forgetting quantitative factors such as land standards for green areas and public facilities, the plan expressed a clear will in a physical sense through the lines it drew on maps.

The new city councils were conscious of the value of the plan that they had inherited and did not fall into the trap of revising it. The existence of the Metropolitan Corporation, presided over by the Major of Barcelona, certainly helped in the early stages. This was in spite of the mistrust of some local councils towards this institution. Thus, a general agreement was reached by the municipalities encompassed in the scope of the PGM to treat it as a very useful tool for the democratic planning project.

Nevertheless, the plan had to be read intelligently, as although its basic principles could be respected, this did not mean freedom did not exist to find more appropriate alternatives depending on the specific moment and place, without this affecting the general coherence (in space or in time) of the PGM.

The nature of the Metropolitan General Plan as a planning instrument established by legislation, along with the precision of the PGM's own content, meant that its development in local plans has required the PGM to be modified on numerous occasions over its lifetime.

This high number of necessary modifications could have led to the belief, for some years now, in the obsolescence of the PGM, but a totally different interpretation is surely more accurate. On the one hand, it can be said that the plan was as a whole of good quality, as it has undergone numerous modifications without its essence being affected. Furthermore, it has been noted that beyond the often unnecessary conflicts between the Generalitat de Catalunya (Autonomous Government of Catalonia) and the local councils, leading to modifications in the PGM, the general management of the project, carried out by the two institutional levels, has been clearly positive: it has given rise to a large number of proposals with specific logics, without questioning the fundamental basis of planning in its area.

This capacity to be modified is surely one of the keys to its prolonged existence. These modifications can be understood as deformations of the plan's general structure that, just as occurs with the structures of buildings, absorb the loads caused by use and so avoid collapse. In the explanation of Barcelona's sequence of planning projects since 1979, it has been necessary to recognize the PGM as the basic reference point.

Localized planning projects: the PERIs

The planning contained in the PGM did not, in spite of its precise nature, exhaust the city's projects; rather the opposite happened. The PGM was a starting point for projects that were being developed, adjusted and also modified. First, let us consider those projects that were based on a close-up vision of space, and that therefore would cover limited areas, but allow considerations of aspects which are not perceptible on a wider scale. Amongst this type of project, the special plans for interior reforms, the PERIs,[1] are especially worthy of note (Figure 7.1).

The interest of Barcelona council in projects of this kind was based on the concern of the new democratic councils for the existing built-up areas in their first years. The period in which the PERIs were most important in municipal policy, especially in the process of drawing up projects and in public debate, was between 1980 and 1986. This municipal democratization was consistent with a concern to improve the existing city for different reasons: the recovery of a city council that represented the interest of citizens led, in the first place, to an increase in the level of self-respect of the citizens as a group. This made this group more sensitive to the needs of improving the urban environment. Second, people who took part

7.1
PERIs (Special Plans of Interior Reform) prepared for the Ciutat Vella (Old Town) in the early 1980s: schemes carried out up to 1998

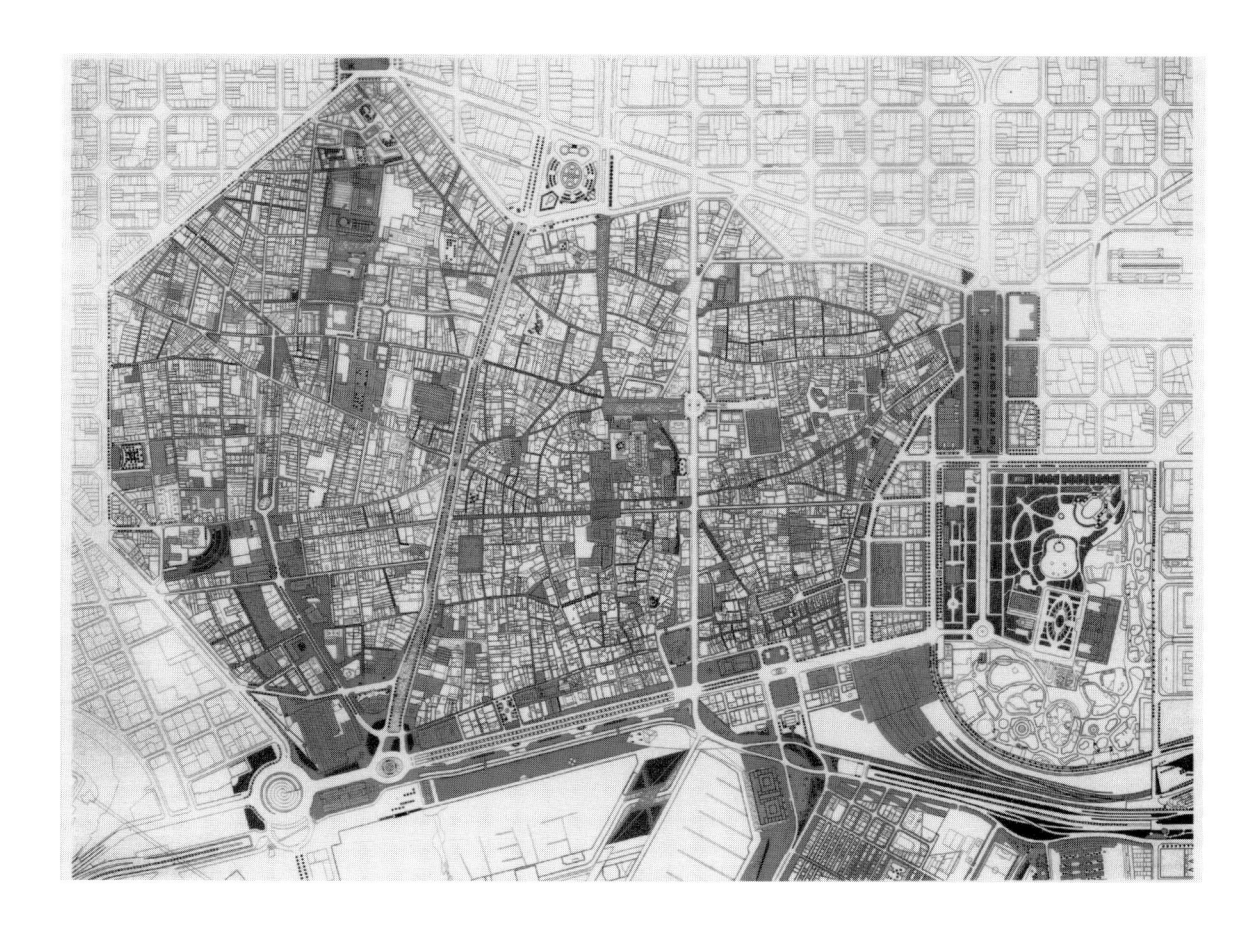

THE PLANNING PROJECT **119**

in the campaigning neighbourhood groups during the last years of the Francoist councils joined the democratic city council as councillors or experts. Finally, it should not be forgotten that at this moment, cultural winds of change within European planning were leading to the defence and recovery of the values of the European city — the Brussels Charter, the proposals of the Krier brothers, Berlin's IBA, etc.

In this context, the municipal concern for the existing city at the beginning of the 1980s took two parallel paths: actions within available spaces — public spaces and purchased sites — and the formulation of plans for interior reform.

The specific planning and improvement actions in the urban fabric with the creation of new squares, parks and public facilities represented an expression of the objectives of the new city council and a demonstration of the efficiency that democracy itself represented. This range of small and medium actions to improve the urban fabric often served as an argument for the discrediting of a plan with obsolete instruments and standards, against a true action-based planning. In spite of the fact that the PGM could not be classified as a plan of this type, it was affected somewhat by this campaign to discredit planning, in the sense that sometimes it was blamed as an obstacle to be overcome. However, a wider perspective demonstrates that without the framework of a general plan, it would have been difficult to carry out many of the actions undertaken.

The PERIs were also relevant to a degree in arriving at this position. This was because they adopted a revisionist attitude, up to a point. The PERIs were instruments that translated the inevitably uniform decisions of the large-scale General Plan to the specific conditions of each area. This objective took on a complex dimension that gave the PERIs an importance that went beyond that of a simple planning instrument.

- First, the PERI was often a medium for the affirmation of the uniqueness of the neighbourhood over the general regulations of the PGM.
- It was also a vehicle that channelled the demands of neighbours, especially for green areas, public facilities and the conservation of spaces and important buildings not considered by the PGM.
- In some cases, it was also an instrument of opposition to specific schemes, basically for roads, proposed or retained by the PGM.

These factors (often quite frequent) gave the PERI a defensive role in the most localized and short-term objectives, against the breadth of scope and long-term vision of the PGM.

For this reason, the PERIs developed in many places not originally foreseen by the PGM, coming sometimes from the demands of neighbourhood associations, who felt that a neighbourhood without a PERI would be forgotten and at a disadvantage, against a plan which was seen as too general.

According to this logic, the PERIs were also an instrument for real planning, compared to the bureaucratic planning of instruments, standards and impacts. Thus, immediate results were often expected. This confidence in the PERI as a rapid solution to all the problems in the urban environment was undoubtedly the origin of many disappointments, which in fact were really the loss of innocence. Citizens finally realized that there are often no easy, quick solutions to the reform of urban areas and that these solutions almost always carry with them a high cost, and resources are always scarce. Thus, over time, the PERIs that have been redefined and readjusted have regained their drier instrumental image.

The fact that the PERIs were often written in the style of 1980s planning is related to the meanings given at the time to the PERIs. This style was especially reflected in the graphic design of the proposals. The proposals were intentionally designed to put forward an ideal urban situation, leading to citizens' objectives that were often merely intuitive. This clearly made them feel closer and more in touch with the general public.

In any case we should note the value of this provision of images, often with distant aspirations, sometimes as reminders of actions which were pending. However, they have always been means to show specific planning objectives in the place concerned, whether to assess, implement or change them.

The implementation of the planned PERI proposals has been varied. The conditions in each place, the viability of the project, and the resources employed in each case have influenced the degree of execution that can be seen today.

The proposals of the PERIs represented a development of the town planning projects of the PGM. It is easy to see the difference between the PGM's map-based plan and the plans resulting when the PERIs were involved. The enrichment of planning and the improvement in the viability of these proposals represents an undeniable advance in the city's overall project. Some proposals have been carried out, some will be carried out, and others will have to be reconsidered (Table 7.1). It should not be forgotten that the planning project is a living project, reflecting the life of the city, and the implementation of these proposals, especially those that affect the urban fabric, frequently require time and successive planning attempts.

As can be deduced from the above, in this section we are referring to a specific type of PERI: those dealing with significant districts or neighbourhoods of the city, plans normally begun at the start of the 1980s. Other PERIs, those of a generally limited scope that had a purely instrumental reach, are not considered here. Others not discussed here include those, more common recently, which remodel large obsolescent areas.

Table 7.1   **Main PERIs**

PERI	Plans and projects[1]	Plans up to 1992[2]	The second renewal[3]
Ciutat Vella			
• Raval*	•	•	
• Sector Oriental	•	•	
• Barceloneta	•	•	
Plans for the Passeig de la Zona Franca	•	•	
Traditional neighbourhoods			
• Gràcia			
– First, more ambitious version	•		
– Second, version of action units	•		
• Sants-Hostafrancs	•	•	
• El Vapor Vell		•	
• Poble Sec		•	
• Sant Andreu (studies not approved)		•	
• Horta		•	
• Prosperitat		•	
Plans for the Carmel	•	•	
Self-construction neighbourhoods			
• Torre Baró	•		
• Vallbona		•	
• Roquetes		•	
• Can Caralleu		•	
• Sant Genís		•	
Latest neighbourhood plans			
• Porta			•
• La Clota			•
• Trinitat Vella			•

* Preceded by the project for the Liceo to the Seminary, which expressed the form of architectural intervention in the urban fabric.

Sources:
(1) Area d'Urbanisme (Ajuntament de Barcelona) (1983) *Plans i projectes per a Barcelona: 1981–1982*, Barcelona: Ajuntament de Barcelona, 297 pp.
(2) Serveis de Planejament Urbanístic (Ajuntament de Barcelona) (1987) *Urbanisme a Barcelona. Plans cap al 92*, Barcelona: Ajuntament de Barcelona, 194 pp.
(3) Ajuntament de Barcelona (1996) *Barcelona. La segona renovació*, Barcelona: Ajuntament de Barcelona, 222 pp.

The major planning projects: the search for general reference points

The more localized planning projects had a leading role in the first stage of democratic planning, in as much as they were assimilated and accepted instruments for the PGM's proposals. In addition, the projects for public space (squares, parks, the improvement of some streets, etc.) were the most obvious signs of the will for tangible action by the new city council in 1980–82.

From 1982, with the formalization of the first ideas of the Olympic project, a new order of actions was proposed. The scale of the city as an overall context reappeared, in which new actions for the city's transformation were conceived. Thus, without ignoring the effect of positive "metastasis"[2] through regeneration of the urban fabric by means of "small-scale" operations for improvement spread around the city, and without also ignoring the objectives for town planning expressed in the "intermediate stage"[3] of the PERIs, from 1982 various projects appeared that could be termed as strategic: strategic because a certain vision for the overall transformation of the city is inherent in them.

These are the "great projects for the city" that complement the piecemeal action of the "plans for the great city".[4] Of special note are the Olympic project, the proposal for areas of new centrality and the road network plan.

The planning project for the Olympic games[5] was the first to take shape from ideas that wished to take advantage of the capacity of urban transformation and improvement implied by an event of this type. The aims were to:

- open the city to the sea;
- distribute spatially the improvements and re-equip the city's sporting facilities;
- promote communication infrastructures, especially the road network.

These objectives expressed the will that the city as a whole should take a great leap forward.

The opening of the city to the sea, with its standard bearer of the new waterfront, the *moll de la Fusta*, under design at the time, required a full-scale operation on the sea frontage, where the city conserved an undeniable potential for urban renewal.

The location of the future Olympic Village on the coast, on the axis of carrer Marina, therefore had a clear strategic intention for the city, because it opened a potential axis of communication with a renewed coastal space. Only an objective of this scope could justify the element of risk of this decision, in a transformation operation that would have a permanent impact.

It should be remembered that, although in the municipality of Barcelona there are not many available pieces of land with sufficient capacity to site 3,500 homes, the metropolitan scope of the Games, reflected in the distribution of sports installations in various municipalities, could have helped to find a location for the village that would not be complicated by existing structures or be limited by general infrastructure (trains, ring roads and sewers). However, the strategic value of the Olympic Village as a factor of general transformation would have been seriously affected, if a metropolitan location had been chosen.

In addition to the Village, the Games areas were distributed within Barcelona municipality in three further Olympic zones. In these three cases, although their scope was uneven and without the dimensions of the Olympic Village, their location took into account the regenerative effects that could occur in each area in a significant part of the city.

In Montjuïc, the objective was to complete the urbanization of the mountain as the most important park of the city. These new installations would strengthen its functional role in the city, and the celebration of the Olympic Games would reinforce its meaning in the citizens' collective imagination. It was quite clear that any spatial planning of the Olympic Games must include Montjuïc.

If in Montjuïc the proposal led to the formation of citizens' space, in the Vall d'Hebron the idea was to turn around the process of formation of a more or less peripheral urban fabric which was occupying the area. The actions would re-equip the area, while its strategic situation in relation to the proposals for the Ronda de Dalt ring road and the Horta tunnel would allow the services that the new sports installations would provide to be more easily accessed.

Thus, the planning in this area was characterized by a vanguardist model that proposed new regulations and even tested new materials for the treatment of urban space. The inclusion of new public facilities such as the reconstruction of the pavilion of the Republic of the Grup d'Artistes i Tècnics Catalans pel Progrés de l'Arquitectura Contemporània (GATCPAC) and the provision of various well-placed sculptures in these new urban spaces was a clear example of the integration of the periphery by means of actions that can be understood as included in the concept of monumentalization.

Less important was the development of the area of Diagonal. This sensibly attempted to incorporate numerous existing sports installations in this area into the programme of the Games, and brought planning order to a confused space at the meeting points of the municipalities of Barcelona, Esplugues and L'Hospitalet. The construction of the Juan Carlos I hotel also reinforced the presence of this area at the western gateway of the city.

The model of the four Olympic areas attempted to take advantage of the beneficial effects of the reforms due to the Games in four unique zones of the city, to become, to a greater or lesser extent, diffusion nuclei for the city's improvement. This model generated another project on the same lines: the areas of new centrality.[6]

This project, prepared before Barcelona's nomination for the Olympic Games in 1986, set out ten areas of the city as proposals for "new centrality", in the sense of spaces in which it was considered opportune to favour a certain concentration of service sector uses and public facilities, in order to create new reference points in the geography of the central areas of the city. It is clear that none of these areas wished to compete with the principal, historic centre of Barcelona, Ciutat Vella and Eixample. Instead, they formed a necessary complement that would articulate and focus the urban fabric, beyond what was possible from the centres of the old incorporated municipalities, which were of a very compact historical nature. The areas of new centrality were not alternative proposals to the centres of Sant Andreu, Horta, Sarrià, etc., whose importance had been reinforced as a result of the decentralization of the municipal administration's districts. Rather the areas of the new centrality (ANCs) provided conditions for the location of new central uses. The only alternatives to these were the Eixample or suburban locations.

The consideration of the Olympic areas as part of this project, made possible by these areas' diverse nature, was in addition, a way of maintaining the sense of the Olympic project (which for timing reasons had to be started before 1986), in case it turned out that Barcelona was not nominated to organize the Games.

The concept of these areas of new centrality can be considered as an evolution of the idea of "directional centres" that the PGM proposed, according to the 1970 Italian model (Figure 7.2). But it gave them more importance in terms of the implantation and impact of the centrality's structure, trying at the same time to redistribute the central uses of the metropolitan area. In Barcelona specifically, the PGM proposed a directional centre in the Renfe-Meridiana district, taking advantage of abandoned railway land, and a second centre, Provençana-Litoral, that was later to receive the name of Diagonal Mar (including with it the relocation of the large MACOSA factory). These two areas were also included in the ANCs proposal.

A total of ten areas were proposed in the project. Two of them were Olympic areas: Vall d'Hebron and Vila Olímpica (the other two, Montjuïc and Diagonal, due to their exclusive emphasis on public facilities were only

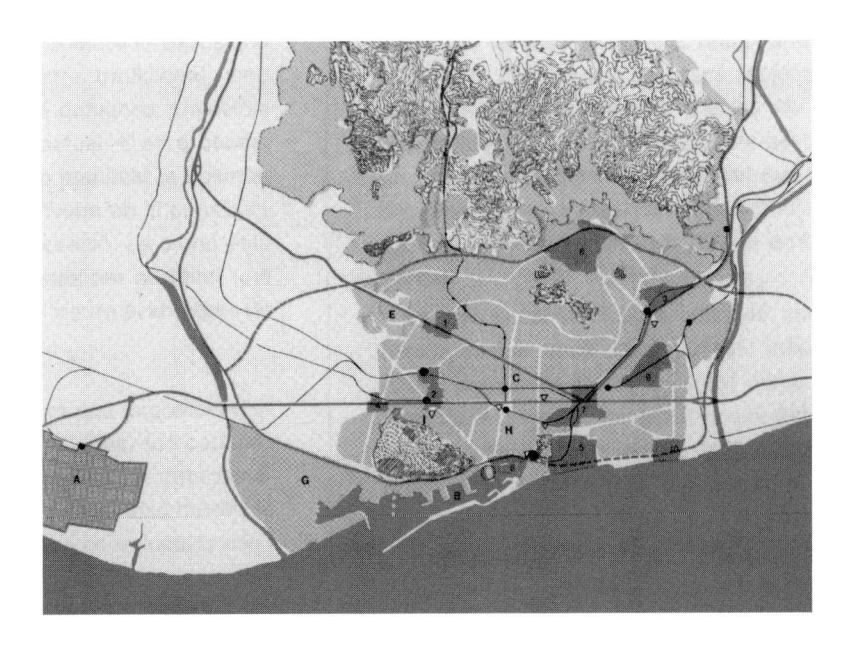

7.2
Areas of New Centrality, schema approved in 1986

added as numbers eleven and twelve). A further two were the above-mentioned directional centres of the PGM. Two more were spaces around major highways, along the axis of Gran Via, around the Glòries and Cerdà squares. Another two were found close to the two central stations of the city's basic railway layout, one already existing (carrer Tarragona, near Sants station), and the other to be opened in the future (la Sagrera). Another area took advantage of the renewal of spaces on the natural axis for extending the tertiary district, on the Diagonal. Finally, Port Vell (the Old Port) was also an area of new centrality, representing the will to integrate it into the city's leisure activity and into the service sector in general (Figures 7.3 and 7.4).

A common factor of these areas was that they all enjoyed spatial conditions that allowed them to take in new types of buildings for the service sector and public facilities, either because sufficient empty space was available or because the use of the space had become obsolete allowing urban renewal. In the majority of cases, the railway and public transport infrastructure, either existing or considered for the future, gave these areas the scope to have a more major function in the city as a whole. Nevertheless, it should also be said that these areas had a diverse content and a variety of different owners and that therefore, the operations and timing of development could be quite variable.

Another aspect is that the use proposed for these areas within the ANC project was not, in any case, in conflict with the uses and building regulations that the PGM permitted. Thus, it can be said that this project was implicit in the PGM, as it could be developed from this base through area-by-area actions. However, this does not reduce the importance of the new centrality areas project. What this project has given is the idea of distributing the new centres throughout the whole city, which undoubtedly increased the sense of every one of these actions. But this is not all. It is also important

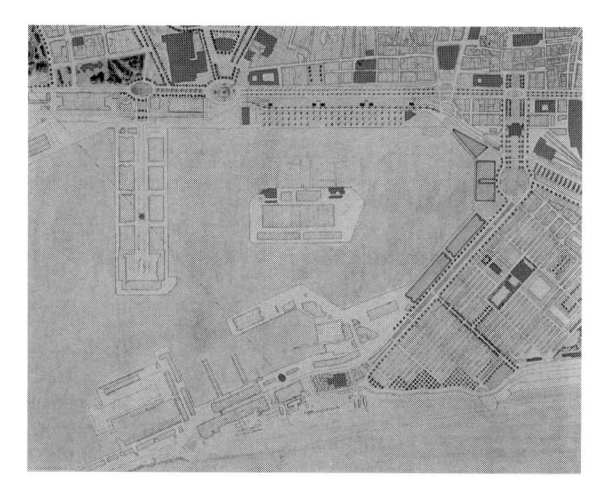

7.3
Scheme for Port Vell (Old Port), mid-1980s (differing greatly from that implemented)

7.4
Further scheme for Port Vell (Old Port), similar to that implemented

to recognize the ANCs project for its notable educational value, in terms of the space of the city as an option for the implantation of activities. The ANCs concept and the project itself have been etched onto the consciousness of both operators and citizens. This has been a very useful tool for the understanding of Barcelona's urban structure.

An area of new centrality that deserves special mention due to its specific nature is Port Vell. It has enjoyed spectacular success among citizens, but a critical attitude from planners and various social commentators and actors.

This criticism has sometimes been unrealistic ("the buildings do not let us see the sea"), but in other cases it has been more conceptual and serious — "a port is by its essence an empty space; full spaces are the city and this should be surrounded by emptiness, with its important role; the introduction of certain uses trivializes such a singular space; the buildings of the port affect the image of Barcelona's sea front". To an extent, it is difficult not to agree with such criticisms and accept them as right. However, in an overall assessment of these actions, one should be aware of what the Port Vell operation represented. It was carried out by a body that was not the city council, and was a factor for improvement and revitalization of all its surrounding area, where in addition extremely interesting urban spaces have taken shape.

Another area in which Barcelona made great efforts to provide a planned reference point beyond that established by the PGM was in the road network. This was very necessary in this case, because amongst the decisions that the city council took in the first years of democracy was the removal of various impacts caused by reservations of land for main roads. We may mention the road from Drassanes to Raval, the Cambó avenue at the eastern sector of Ciutat Vella, or the via O in Gràcia and the continuation of the first ring road in Horta, Sant Andreu and Sant Martí. The latter was implicitly stopped by the demolition of half a viaduct of the final stretch, and by the connection of the rest with the Rovira tunnel.

In order to debate the criteria for a new road network proposal, the council and the metropolitan corporation organized in 1984 a seminar,[7] with the participation of experts, specialists and politicians. Its conclusions reinforced the thesis of the interdependence between the road network and the urban fabric, the role of the road network as an integrating factor in urban space, and the value of arterial elements as distributors that strengthened the secondary network rather than as through roads.

In this seminar, the work of the Barcelona Highways Plan[8] was presented. This plan became a basic reference point for highway and planning actions in the city. It functioned as an internal reference point because, although it moderated the PGM road network proposal and gave greater urbanity to it by introducing various changes, no modification to the PGM was made to increase its degree of official status. The framework of the PGM, with large reservations of land, gave space for the plan's highway proposals. The adaptation of road networks to the urban fabric, one of its principles, would be carried out gradually, as projects were carried out in the corresponding areas, at which time the general plan would be modified. This technique, later to be applied in other cases (the Coastal Plan, Collserola, etc.), allowed the combination of the necessary stability of the general planning framework with a fruitful process of successive advances made by the planning projects themselves.

The metropolitan planning projects

As has been shown, in 1974 the Barcelona Metropolitan Corporation (CMB) was created with powers of urban action within the scope of the PGM. This institution was abolished in 1987 by the Catalan Parliament with the approval of the so-called planning laws.

During its years of existence, the CMB extrapolated at a metropolitan level the planning ideas and criteria of the new democratic councils. Especially interesting contributions came from Barcelona city council, but there were valuable ones from smaller councils, too.

It should be remembered that the governing bodies of the CMB were made up of representatives from the municipalities (generally direct representatives or in some cases provincial councillors). It was the mayor of Barcelona who presided over the years of this institution's existence.

It is important to note one interesting aspect of the nature and role of this body. The CMB acted as the planning commission for everything related to the approval of partial and special municipal plans. These planning powers obliged it to exercise a type of self-control, as in fact it was the municipalities of the metropolitan area that together considered the value of the plans presented by each one of them.

Undoubtedly, throughout its existence, the norms of behaviour in the municipalities, over and above the style of some mayors, managed to consolidate the sense of belonging to a single urban space. In spite of the break-up of the CMB and its fragmentation into specialist bodies, this has allowed the maintenance of a metropolitan nexus through the creation of a voluntary association of municipalities.

It should be pointed out that the annual availability of an investment fund from the general budget of the Spanish state contributed to this maintenance of formulas for metropolitan cooperation. These quantities have been received on the basis that the area's municipalities as a whole constituted a metropolitan reality in terms of population and intensity comparable with what in Madrid is a single municipality. It was therefore necessary to find a way to even out these contributions from the state per inhabitant by means of inter-regional compensation. However, once the CMB was dissolved, these funds ended up in each municipality and it is now the municipalities that voluntarily contribute towards the association of municipalities. In addition to acting as a regulating body for municipal urban planning, the CMB also carried out its own planning programme. The principal reference point of the action was naturally the PGM, which included precisely the CMB area.

The planning position of the CMB was to consider the scope of the real city. Without ignoring its true extension, which should probably be seen as larger, it was an area with more planning consistency than that of each municipality. This scope required and permitted other ways of focusing. Nevertheless, municipal realities should not be forgotten, realities which had been logically strengthened by democratic representation in the local councils.

The two fronts of the PGM's urban development projects (area-based and city- or municipality-wide, as discussed above for Barcelona), have also featured to a different extent in the metropolitan area's municipalities. The PERIs, with similar meanings to those noted for Barcelona, have been present in all the municipalities: city-wide projects have also been present in diverse forms in the most important municipalities.

The CMB had to develop its planning activities with special consideration for the initiatives of each municipality. Some municipalities with the lowest level of technical capacity requested the cooperation of the CMB in specific cases.

In this context the CMB focused its planning contribution in three principal ways:

- the creation of metropolitan parks;[9]
- the construction of road links;[10]
- metropolitan planning projects.[11]

The creation of metropolitan parks corresponded rather well to an extension to the metropolitan sphere of the policy to create new parks, that Barcelona council introduced during its early years (Escorxador, Espanya Industrial, la Creueta del Coll, etc.). In the first park building programme started by the CMB in 1982 (Torreblanca, Besòs, Can Solei, Torrerroja, etc.), the reference point of the still-to-be-finished actions of Barcelona took a very prominent role. The metropolitan park programme lasted throughout CMB's existence, with parks including les Planes, Can Mercader, Central de Sant Vicens, del Molinet, litoral del Besòs, etc., and has continued into the present with Area Metropolitana de Barcelona Mancomunitat de Municipis (successor body to the CMB) creating Les Aigües, Can Massot, el Turonet, la Muntanyeta, Parc Nou, etc.

The park building programme of the metropolitan institutions (initially termed metropolitan parks in reference to the developing agency) set out literally to monumentalize the periphery. Clearly, the periphery in the metropolitan area was especially extensive, including urban areas that have never reached the level of quality that the word periphery implies. In this sense, the CMB's action on parks was not so much based on dealing with parks of a particular size, but rather of concentrating principally on the distribution of urban scale parks in the metropolitan area beyond the limits of Barcelona, with the intention of extending the "monumentalization" to the whole periphery of the metropolitan area.

The creation of road links was centred on building inter-municipal highways, fundamental for the improvement of a metropolitan road network that, until 1976, had been developed in a very piecemeal manner. It should be pointed out that this was a task particularly appropriate to a metropolitan administration, as they were schemes that exceeded the reach of municipal actions, but which at the same time did not form part of networks controlled by higher level authorities. The construction of some bridges, such as over the river Besòs between Sant Adrià and Badalona, over the river Ripoll between Ripollet and Cerdanyola or over the ravine Els Gorgs in the Parc Tecnològic del Vallès (Vallès Technological Park), obviously responds to the need to connect road networks. However, it does not ignore the potential of bridges to be transformed into monuments as potent and unique formal reference points.

As far as wider planning activities are concerned, mention should be given to two especially significant proposals: the Coastal Plan and the Special Plan for Collserola.

Those parts of the metropolitan area with a clear geographical definition were reordered within the framework of the General Plan, studied with the flexibility that a large area allows: the 40 km of coast and the 6,500 hectares of the Collserola massif designated as forest park by the PGM.

These projects were followed by others such as the planning of rural space or the integrated planning of space and infrastructure that was tested for the lower river valley and delta of the Llobregat. The disappearance of the CMB led these latter studies to have a weaker formalization and impact. However, a notable clarity can be seen in some of these matters, which have still not been totally resolved some ten years later.

The Coastal Plan

As has already been said, the coast within the limits of the PGM is some 40 km long. Aside from the different land uses and conditions, its proximity to the sea gave this strip an undeniable uniqueness and a clear potential for improvement.

The drawing-up of the Coastal Plan[12] was begun in 1983, at the same time as:

- the city had been rediscovered as a leisure space, with a growing demand for different possibilities of urban and metropolitan leisure;
- the new sewer outfalls gave the northern stretch of coastline clean waters;

- the possibilities for the regular regeneration of beaches by bringing in marine sands was becoming better known and its probable unsustainability was not suspected;
- the initial project for the Olympic Games (Cuyàs Report 1982) proposed significant interventions along the coastal strip, especially the Olympic Village, as an operation for the recovery of coastal space by means of a major scheme of residential remodelling.

In this context, the Coastal Plan was in the first place an instrument for disseminating the potential of the metropolitan coast.

Initially, the plan was intended to have significantly more management powers. In a moment of strong dynamism for action by the CMB, the possibility was considered that the coastal strip should be a territorial unit of metropolitan management. It was soon seen that this was impossible. In addition to the size of the area covered, which required many smaller actions, the plan included urban spaces that were too significant for some of the municipalities to give up direct management over these spaces in favour of a hypothetical single metropolitan management.

Considered in strict planning terms, there were also signs that the largest councils carried most weight in decisions. The clearest example was the planning of the Olympic Village. It was drawn up in parallel to the Coastal Plan, but by another team, and finally adopted criteria that were clearly different in terms of the characteristics of the proposed development. However, consistency with some basic criteria was kept (the ring road, the Marina axis, etc.). All of this represented a significant reduction in the initial objectives of the plan, but it was by no means an attempt that failed.

First, the plan had an epistemological role, as it facilitated the discovery of metropolitan coastal territory and expressed in images its possibilities for change.

In relation to its most immediate functioning, the Coastal Plan:

- constructed an overall discourse on which were built the specific transformation proposals of various key points along the marine front such as Barrau in Montgat and Unió Vidriera, Campsa and Cros in Badalona;
- explained the boundaries and contents of the various possible operations needed for a process of urban coastal recovery.

Thus, I believe that after the Coastal Plan a different idea was developed of the reality and potential for the metropolitan coastal strip, and that this undoubtedly had a positive influence on the on-going process of the transformation of this space (Figure 7.5).

Reform of the Ciutadella Park

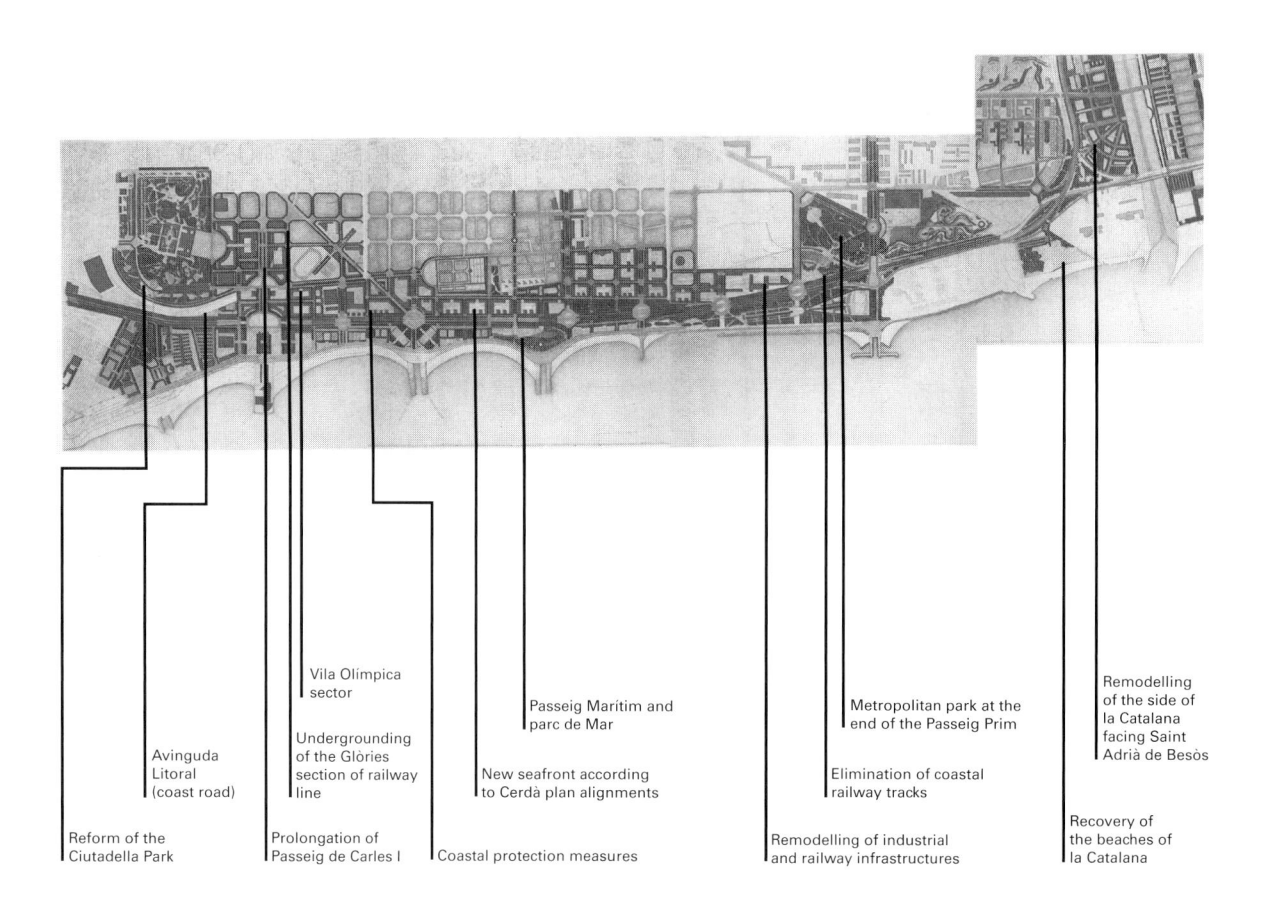

Avinguda Litoral (coast road)

Vila Olímpica sector

Undergrounding of the Glòries section of railway line

Prolongation of Passeig de Carles I

Passeig Marítim and parc de Mar

New seafront according to Cerdà plan alignments

Coastal protection measures

Metropolitan park at the end of the Passeig Prim

Elimination of coastal railway tracks

Remodelling of industrial and railway infrastructures

Remodelling of the side of la Catalana facing Saint Adrià de Besòs

Recovery of the beaches of la Catalana

7.5
Section of the Pla de Costes (Coastal Plan), prepared by Metropolitan Corporation, approved 1987

The Collserola Project

The zones of the Collserola range of hills classified in general plans as forest park occupy approximately 6,500 hectares. The areas included in the Collserola geographical area constitute about 20% of the area included in the PGM.

The two factors of size and morphology demanded a specialized treatment for the area, which is also extremely important in the traditional leisure activities of the municipalities which surround it.

A stage prior to this project was a process for strengthening the identity of this area, which peaked in 1982 when a series of Collserola workshops and conferences were held. This process was necessary in order to become familiar with the scale of the range of hills, which is neither that of Montseny — also a metropolitan forest park and which already had a special plan in 1976 — nor that of Montjuïc, which is the great city park of Barcelona.

Tailoring the project to address the scale of Collserola was not achieved immediately. The metropolitan government was especially concerned with the image of intense large-scale urban transformation which was to be carried out in order to prepare the city for the Olympic Games (for which Barcelona was still only a candidate), and it was understood that Collserola had to be converted into a park using similar methods. This led to the suggestion, made at one point, that Ricardo Bofill's team should be in charge of designing the project for Collserola.

Once the nature of the problem was understood, the planning of Collserola was tackled progressively, as agreements were looked for between the differing degrees of interest in it from a naturalist standpoint, for the conservation and the improvement of the traditional landscape, and for the conditioning of a network of routes and places to be used more intensely by the citizens.

I believe that the Plan for Collserola[13] provides a vision which complemented that of the PGM. While in the latter, Collserola was the central void which was respected by the urban environment, in the Collserola Special Plan (PEC), it was the new observatory from which to evaluate the fabric and the new structures located in the urban areas which surrounded it. In this way a document was produced which provided different visions of the categories of forest land, the surrounding urban ring or crown, and the various highway elements (the Horta tunnel, the cornice route, etc.) which led to some of the alternatives to the PGM's allocations.

In the case of the Collserola Special Plan (PEC) a form of coexistence was created between allocations which did not coincide. Some of the proposals of the PEC (such as the unifying of forest classifications) stimulated the introduction of changes to the PGM. Others coexisted, such as those aspects related to the routes mentioned above; aspects of the PGM referring to routes remained in effect while those in the PEC were considered suggestions for the future, although they were backed up by the ideas in the Special Plan.

I believe that this was a good decision, defended by the director general of planning of the Generalitat, not to carry out any modifications to the PGM which were not specifically required for the immediate work at hand, and to accept the different proposals that formed part of what could be the park strategy, which showed a direction for change, but did not create any different, long-term, binding planning norms.

In comparison to the Coastal Plan, in Collserola it was possible to define an area for unitary management and the creation of a metropolitan entity which was specifically responsible for carrying this out, the Patronat Metropolità de Collserola. This was possible because of its size, its peripheral position compared to the coastal zones at the centre of each municipality, and its morphological distinctness. These two different approaches, in considering two different territorial projects in the same region, demonstrate the need for an understanding of the context of each project and of their significance for the municipalities, which the CMB's planning actions required.

The restoration and improvement of buildings

One defining characteristic of the spaces in our cities is that the land is divided into two complementary and mutually exclusive categories: the network of public spaces, and the plots of land between these. The former represent the spaces through and along which we can move, and via which we can access the buildings. From these spaces the buildings receive light and ventilation, and from them we can see the facades and shapes of the buildings. That is, we perceive the architecture of the city from these spaces. The spaces which these public areas divide up and define are basically the plots of land that the buildings occupy. In contrast to the public spaces, these plots of land make up the filling of urban space, and they are characterized by private use, although some buildings are in fact public.

The degree of consolidation of the urban texture made up by the network of public spaces and the plots of land varies according to the different areas of the city. Some parts of the city can be said to be completely consolidated, as in the Eixample. In other parts it is somewhat less consolidated, to the extent that the PERIs aim to change some blocks and streets to some extent, as in Ciutat Vella and to a lesser degree in other PERIs which have been mentioned. Finally, there are areas which, due to the possibility of a considerable degree of remodelling, should not be considered as consolidated, as is the case of the areas dealt with in the following section. In the non-consolidated urban areas in which substantial changes to the order of the area are expected, the characteristics of the overall renovation of the area, although normally carried out sequentially, are determined by new projects.

In the other parts of the city, which without doubt are the majority, the proposals for renovation vary between those that are aimed at public spaces and those that are aimed at the plots which are built-up.

The remodelling of many of the public spaces that already exist in the city, along with the creation of new spaces resulting from planning, has been a central theme in the transformation of Barcelona.[14]

This section concentrates on the actions which were designed to restore and improve the greater part of the built-up city. These actions should be seen as forming part of Barcelona's planning project (in that it supports the maintenance of the characteristic features of the buildings and image of Barcelona), but are not what we would normally consider to be part of a conventional planning project (for example the PERIs project), which would normally go no further than the graphic representation of the desired urban order.

It should be added that, in spite of the attention which it has been possible to pay to the sum of the city's buildings in these years of considerable urban transformation, the situation of the buildings does not equal the quality that has been reached in public spaces. The landlord and tenant law which was in force up to 1995 is responsible, to a large extent, for this situation. This is particularly true in a city like Barcelona where the percentage of dwellings which are rented, more than 30%, is much higher than the average for other Spanish cities.

By definition, the project for the process of restoring and improving the city's buildings affects most directly spaces, mainly under private ownership, where any action must be carried out in collaboration with the owners.

In general we can distinguish three types of actions which represent different means and degrees of involvement of the owners:

- campaigns and programmes to encourage the private owner to act, where the techniques employed are based on persuasion;
- rules and regulations which the private owner is obliged to follow;
- direct action by the administration, substituting the action of the private owner.

The campaigns have tried to combine the creation of a climate of participation with the provision of economic incentives for the actions of private owners. The prime campaign of this type has been "Barcelona posa't guapa" (Barcelona, look smart), which has been underway since 1986 and which has been responsible for the creation of the Agència del Paisatge Urbà (Urban Landscape Agency).

As is clear from the title, the campaign is directed specifically at improving the existing image of the city. The argument was based on the concept of urban landscape, that is to say, it was centred on the qualitative value of the images which the city provides us with.

Most of the iconographic capacity of the city resides in its buildings, or to be more precise in their outer appearance. So, with the incentive of one year's saving on the property tax on the building, it was mainly the restoration of facades and outer walls that was promoted, along with some other improvements in varied aspects of the urban landscape such as gardens, fences and walls, signs, etc.

The response to this campaign has been considerable, to the point where it has become an example which it is impossible to avoid when we talk of citizen involvement, or collaboration between the public and private sectors. The restoration of the facades of more than 3,700 buildings is clearly noticeable around the city.

Although this campaign was directed at the most superficial aspects of the buildings in the city, the outer appearance of buildings is a public manifestation of the private domain of the city, and not insignificant. In such cases the most important thing is to establish a dynamic of improvement. When the owners of a building improve the exterior, it will probably be accompanied by a heightened attention to the maintenance and improvement of other aspects, because the building has moved up in the estimation of the owner and users. On the other hand, the public announcement of every improvement has no doubt been an important factor in the improvement of nearby buildings, and in this way there is a certain multiplier effect in the campaign.

The development of the programme "Barcelona posa't guapa" has without doubt created an interesting model which the Urban Landscape Agency has continued to extend to cover other components of buildings in later campaigns: antennas, solar energy panels, etc.

Other campaigns have been aimed specifically at recovering the necessary conditions for the profitable use of buildings, through subsidies to private owners for the restoration of buildings.

The restoration of existing buildings is an idea which took on new life in the early 1980s. The waste caused by the premature demolition of buildings, together with the prolonged economic recession and a re-evaluation of the collective memory of the city, backed the ideology of conserving the city's buildings and putting them to good use.

The first legislative action taken by the Spanish state in an attempt to encourage building restoration was in 1983, followed by successive laws in the years 1987, 1991 and 1993. It is also worth remembering that the metropolitan corporation approved in 1985 the metropolitan regulations for rehabilitation, which adapt the regulations of the PGM to the specific terms of rehabilitation.

Within the framework of the subsidies laid out in successive decrees, buildings in all parts of the city have been rehabilitated, although this has not been enough to bring the overall condition of buildings in the city up to date. On the other hand, in some projects for the overall restoration of private buildings, it has been suggested that subsidies should depend on the subsequent use of the restored building. In any case the restoration of buildings in the consolidated areas of the city provides support for the survival of those neighbourhoods and this always has general value.

In Ciutat Vella there has been a large amount of rehabilitation by private owners. In 1986 the Àrea de Rehabilitació Integrada (ARI — Area for Integrated Restoration) was approved in Ciutat Vella, which made it possible to obtain loans for restoration work, at rates of interest considerably lower than market rates.

Since 1988 private rehabilitation has grown. From 1990 this has been made easier by the creation of the Office for the Rehabilitation of Ciutat Vella. However, 80% of the investment in private restoration has been carried out on the basis of agreements between the Spanish state, the Generalitat and the city council which were signed in 1994, based on the 1993 decree. These agreements included the setting up of preferential rehabilitation zones where subsidies were concentrated. These zones varied every two years, in such a way that in six years the whole of the area of Ciutat Vella to be included would be covered.

Private rehabilitation work between 1988 and 1997 accounted for a total private investment of more than 14,000 million pesetas and subsidies in the form of grants of 3,150 million pesetas, affecting 14,800 commercial premises and dwellings, a figure that represents more than 20% of the private buildings in the district.

In relation to the regulations, it should first be pointed out that the PGM itself has some specific effects on the conservation and improvement of the buildings in the city. The main objective of the PGM in the consolidated areas of the city was to control population density, and with this goal in mind it reduced the possibilities for the construction of buildings in the city which had been permitted up to then. In those areas of the city where the layout of the roads determines planning, which is the majority of the city, the possibility for the construction of buildings was reduced by several floors.

This had several different effects over time. Due to unsatisfactory control of the suspension of licences, applications for licences in accordance with the previous regulations increased massively between 1974 and 1976 while the general plan was being processed. These licences became effective over the next few years, which delayed the effect of relieving population density in the city, and caused a shortfall in the number of applications for new licences in the following years.

Secondly, all those buildings with a volume which does not correspond to that set out in the PGM, have the possibility to remain in this situation for a long time, since replacing them would lead to a loss of part of the floor space that can be built on. It could be said that most of the fabric of Barcelona is fairly compact, with only a few empty plots, and that the buildings in the city have a strong tendency to remain as they are.

Within this fabric the Eixample has a special place. The city council commissioned a study from the Laboratori d'Urbanisme (Planning Laboratory), which gave the material for an exhibition at the end of 1983 in the Casa Elizalde, and which marked the beginning of special consideration for this area of the city.[15]

The Eixample is an area with a very regular structure that clearly defines the structure of public spaces (abundant roads and sparse green areas), and where the use of a PERI did not make much sense as the means of remodelling the area. However, it was clear that the Eixample required a much more specific treatment than that which was possible within the framework of the PGM.

The solution was the Ordenança per a la rehabilitació de l'Eixample (The Rehabilitation of the Eixample Bylaw), passed in 1986. This bylaw has defined the area of the Conjunt Especial de l'Eixample (Eixample Special Area) and within it, another smaller area which is called the Sector de Conservació (Conservation Sector).

One of the objectives of the bylaw is the opening out of the area by restricting the occupation of the interior patios of the blocks by new buildings, as a first step towards freeing these spaces for community or public use. It is certainly true that the survival of the Eixample as a residential area in the face of strong competition from the rest of the metropolitan area requires an improvement in conditions, and particularly an increase in green spaces. Projects for the opening up of the patios as a viable option for the improvement of the area were already outlined in the PGM. This was complemented by the bylaw, extended by a more ambitious scheme in 1994, and by the actions of Proeixample, a management body set up to promote and encourage improvements in this area of the city (Figure 7.6).

Another basic aim of the bylaw is the protection of the Eixample as a part of the city's architecture. It does this through the introduction of regulations for the facades of new buildings, through heavy restrictions on demolition within the conservation sector, and also through the creation of a consultative body, the Comissió de Manteniment i Millora de l'Eixample (Eixample Improvement and Maintenance Commission), which has as its remit to report on any developments which affect the space that comprises the Eixample. This commission is, due to its composition, a meeting point for those involved in architecture and property, and the city council, and it is therefore a suitable mechanism for improving the city's architecture.

In fact, architecture is what ultimately gives shape to the volumes which are governed by planning, so the quality of the city depends not just on planning but to a large extent on architecture as well. However, in the same way as public administration has full responsibility for planning, so architecture (except in public buildings) is mostly the responsibility of private operators; developers, architects and users are together the people who are responsible for the architecture which defines the urban volumes contained within the public spaces.

However, architecture is a substantial component of the city project and so the city council should (as indeed it has) aim to improve the architectural quality of buildings.

In the first place, this is done through the preservation of valuable architectural heritage. To this end the general regulations for the protection of the artistic-historic heritage of the city of Barcelona were drawn up and approved. They contain lists of buildings which are subjected to varying degrees of protection. These regulations also include the Eixample as an area to protect, and include, in its entirety, the specific bylaw for its protection and improvement which was mentioned previously.

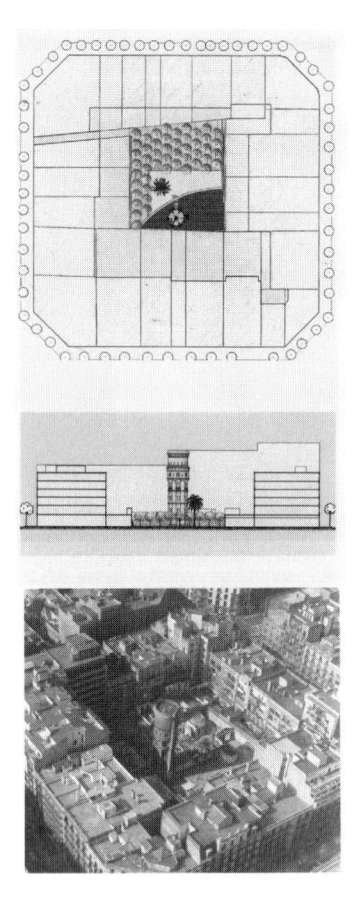

7.6
Pati de les Aigues, plan and aerial view, the first patio freed for public use

Second, support for architectural quality is given by the establishment of regulations for the form of buildings, as is also set out in the bylaw for the restoration and improvement of the Eixample, an important instance of this type of document. This is a complex form of action, in which the decisions are always open to argument. That is why the Commission is a necessary complement for any regulations of this type, both because it can adapt the application of criteria, and because it can function as a place for the transmission of cultural criteria between competing sensibilities.

Third, there is a type of action which it is difficult to regulate and which must be used with prudence. This is to convince the important property developers to work with good architects. This type of action, which has also given noticeable results in Barcelona, would not be possible without a fourth type of action, which is the improvement of the quality of architecture in the public domain.

The buildings which are commissioned by the public administration, whether they are buildings for public facilities of some kind or housing, also form part of the urban make-up of the city and they are quite singular. It is clear that the architecture of public buildings has to set an example, and that public architecture, particularly that which is commissioned by the city council but also by other administrations, has throughout these years maintained a fairly high level of quality.

If we can take for granted the quality of public buildings (whether for public offices or for facilities), given their symbolic value, the quality has also been quite high in the field of public housing. It can certainly be claimed that it has been of reasonable quality in spite of the limitations which have been placed on it by budget restrictions, which are greater than those placed on private housing.

For these reasons the direct physical intervention of the city council and other administrations has had an important effect on the restoration and improvement of the fabric of the city, especially due to its value as a reference point for private developers, who are more numerous and diverse.

When it comes to new housing projects commissioned by public administrations, we must mention the 1,700 dwellings built in Ciutat Vella, as well as 420 dwellings completely restored by public projects. Within these actions, due to their size, the land previously occupied by La Maquinista and that of Avinguda Icària stand out, both of them located in Barceloneta. Outside Ciutat Vella, the promotion of housing to be let to young people near the Ronda de Dalt ring road must be mentioned.

Most public housing has been made available in order to substitute old large housing estates — Baró de Viver, Eduard Aunós, Vivendes del Governador, etc. The rest of the actions, which complement or form part of the consolidated city fabric, certainly have not managed to cause important quantitative repercussions in the city as a whole, mainly due to the shortage of land available. Nonetheless, the goal of architectural quality in these projects should be emphasized.

Recycling urban space: urban transformation projects after 1992

The year 1992 is a key reference point in the sequence of the planning project for Barcelona. It can certainly be said on the one hand that the planning goals for the city continued to be the same, but on the other hand it is easy to see that the celebration of the Olympic Games in the city represented a clear change in the stage of the process. This change came about smoothly, given that after the date mentioned many public and private works were completed which had been conceived and started beforehand, in particular those which were related to the project of new central areas: Teatre Nacional (National Theatre of Catalonia), L'Illa (a shopping centre), el Port Vell (the Old Port), etc. These can be understood as part of the transformation of the city which had 1992 as its point of reference.

In parallel with the conclusion of these processes, the planning project for Barcelona started to incorporate new aims and to adapt its progress to the new circumstances which, as we moved away from the magic date of the Games, became more noticeable.

What has been called the "Second Renovation" was being developed, or as others saw it the Third, with the supposition that you could identify in the previous period two stages: that before the Games became the main horizon, and the period in which the project was oriented towards the date of the Games.

The main changes in the context were the disappearance of a deadline for the work, and the public debt brought about by the previous actions, accompanied by restrictions in the drawing up of the budgets of the different administrations, due to the programme for a common European monetary policy. On another level, the increase in the international prestige of the city caused by the recognition of its management capacity must also be mentioned as a change, along with its recognition as a place of opportunity for international investors and also as a place with privileged conditions for the quality of life.

Using the advantages of this new position meant, in the first instance, giving the necessary attention to the main communication infrastructures with Europe and the rest of the world: the airport, the port and the planned TGV, and this had to be done in a context which did not favour public spending.

On the other hand, it should also be noted that in the years after 1992 the city was becoming more aware of the evolution of other circumstances brought about by its more immediate territorial situation.

First, there was the fact that land to extend the city had run out. In these years we see the appearance of the initiative to develop the Diagonal Mar district, the last piece of land available and in fact almost the only one which the PGM had considered as urbanizable land within the municipal limits of Barcelona. Second, there was the loss of population occurring since 1985. This also affects other municipal areas around Barcelona and is explained, fundamentally, by the absence of space to expand. It also makes clear the competition between metropolitan areas for the location of housing or other activities.

It is surely for all of these reasons that, after 1992, some criteria which had been in the background during the previous period became powerful in the continuous reformulation of the project for the city.

First, the city council understood that it was particularly convenient in these circumstances to make use of private developers as a means of transformation. Despite the recession of 1993, private developers maintained an important presence in the city, encouraged by the image of improved quality of the city. Some of these operators (Travelstead, Kepro, etc.) went bankrupt, but they were substituted by others in the projects which were underway.

Less explicit was the aim of recycling urban land, against the trend to urbanize new land in the metropolitan context. However, this aim is amongst the principal elements of this period.

The encouragement of housing, and especially of housing at affordable prices, gained ground as an objective. The city should be first and foremost an inhabited space, without prejudicing the mixture of uses characteristic of the make-up of the city.

The most important projects from this period were geographically in the north-east of the city, mainly in the districts of Sant Martí and Sant Andreu. This is the part of the city where the potential for transformation was greatest, where there were old projects pending (the opening up of the Diagonal is a particularly clear example), where new transforming factors were expected (the TGV station) and the Olympic Village and the new coastal front began to actively express themselves, as a demonstration of the change that was possible.

In this area, five main projects were formulated which, in spite of the fact that all shared the abovementioned characteristics, had different features. Because of their importance for the future of the city, they deserve individual consideration.

Diagonal-Poblenou[16]

This is a transformation project which was already anticipated in the PGM in 1976. In fact it should be remembered that the continuation of the Diagonal has been present in all the plans for Barcelona since Cerdà. It consists of the opening of a 12 km long road in a straight line all the way to the sea and the remodelling of the obsolete areas which were affected by its route. Work on this project began in 1988, even though it had a logic that was different from other contemporary projects which were targeted at the events of 1992.

It is a project based on public initiative, but it includes the participation of (by their own free will or through coercion) the owners of the land and private developers, through the use of the management tools which planning legislation makes available. It is therefore a long process which has inevitably needed a period for maturation, and which during the year 1998 achieved the 100-year old objective of opening the route of the Diagonal to the sea. This fact will without doubt be a dynamic factor in the process of remodelling the surrounding plots, which have been split into several different units in order to make the conclusion of the project easier.

Sant Andreu-La Sagrera[17]

This project was not proposed in the PGM, since the plan, with its perspective of the time, maintained intact all urban industrial areas. However, this project had been suggested within the proposal for areas of new centrality, although in that proposal it was on a much smaller scale.

Two new circumstances gave rise to the important changes to the ideas contained in the areas of new centrality: the relocation of several large industries (in particular La Maquinista, which occupied 25 hectares) and of the military barracks to outside Barcelona, and the plans to site the TGV station at La Sagrera (Figure 7.7).

The project had an interesting forerunner in the proposal made by Norman Foster (1993), commissioned by a private developer. This proposal, which was something of a fantasy and not very viable, proposed the creation of an artificial river from Trinitat to Glòries. However, the idea of the river favoured the proposal of a park along its banks, which is the key element in the plan now proposed for the area.

Apart from the park, the project sets out an important inter-modal transport centre with the TGV station, and it adjusts the road system for the area, so as to resolve the access problems and define the basic structure of the 240 hectare district.

7.7
Aerial view of proposed Sagrera development (including high-speed train station and park)

Similarly, the project proposes a process of urban renovation of considerable size, spread between several areas. Those areas consisting of large sites are already the object of plans by private developers. Those with fragmented sites, and which are partially occupied, will need a period of maturation. In some cases the intervention of public operators will be necessary in order to get the projects moving.

Front Marítim
This can be considered as the continuation of the renovation of the sea front which was started with the Olympic Village. However, the Olympic Village had not been proposed in the PGM, whilst Front Maritim had. The probable abandonment of the land by Catalana de Gas allowed, even in 1974, the suggestion that the land should be integrated into the urban area.

The project consists of the restructuring of the urban sea front, through a new area consisting mainly of housing. The land was to a large extent in public ownership, and the planning project was carried out by the City Council via a competition for ideas, as to which would be the most suitable configuration for the arrival of the Cerdà grid to the sea front. A reinterpretation of the blocks in the Eixample won, following the lines of three blocks near the Olympic Village, in a project initially named Eixample Marítim.

Once the plan had been established, the different plots were sold by auction to private developers, to implement the plan and construct the buildings.

Diagonal Mar
In this case we are dealing with a project which was private from the outset. The land belonged to a large industrial company (MACOSA) and the IMPU (Institut Municipal per a la Promocio Urbanistica i els Jocs Olimpics) and it was bought at a reasonably high price during the euphoria of 1992 by a developer who wanted (in agreement with the proposal in the PGM) to construct a tertiary centre.

The recession in the office market and no doubt the price paid for the land caused the project to fail, and the land was acquired by a second developer (Hines) which, following different criteria, wanted to develop a housing project.

In this case the city council was open to the logic of the developers (even though this logic proved to be quite changeable in a short period of time), along the lines of facilitating interesting investments for the renovation of the city and of the sea front in particular. Similarly, in this case a type of urban layout was accepted which breaks considerably with the style of the configuration of the rest of the sea front in this quarter of the city. It is a disputed and disputable plan, but it should be pointed out that an end piece to the layout of the city such as this is, allows an experience of this type, which can provide us with different and hitherto unexperienced ways to relate public spaces and buildings.

The Diagonal Mar project, along with that of Sant Andreu-La Sagrera (Sector Maquinista), proposes a large-scale shopping centre. These shopping centres were the important parts of the programmes drawn up by the private developers for these areas. These two projects formed part of the list of areas of new centrality. In the same way several more of these areas (Glòries, Diagonal-Sarrià, Olympic Village, etc.) have incorporated shopping centres which have opened their doors during this period. Within the philosophy of development of the city of Barcelona since 1976, these commercial centres which are integrated into the urban fabric are valued very positively as undeniable factors of centrality in various places around the city. At the same time, isolated commercial centres on the outskirts of the city are strongly disliked as inevitably destructive of the commercial make-up of the urban areas.

Industrial sectors

Finally, we should just mention a project which has not yet been completely formalized, which aims to renovate the industrial sectors integrated in the fabric of Barcelona.[18]

These areas, and especially those in Poblenou, have suffered a process of loss of activity and at the moment they are being maintained under the hope of an increase in the price of the land so that they can switch to a use with a higher demand — housing. As can easily be understood, this situation does not help the natural renovation of the area, which on the other hand is limited due to the archaic definition of industrial use contained in the PGM.

The configuration of the industrial sectors of Poblenou by the Cerdà network of roads facilitates the partial evolution of the area by blocks or by moderate sized plots (Figure 7.8).

The aim of the project being studied is to facilitate the gradual transformation of industrial zones towards a mixed composition, in which economic activities continue to occupy most of the space but without excluding a certain proportion of housing (Figure 7.9).

The key is to find the exact point that favours acts of renovation, without the new dynamic leading to the displacement of the activities of the weakest inhabitants.

This process of urban renovation could be especially interesting, and different from the other projects discussed here. The absence of any final formed image beyond the network of roads gives the process an openness which can favour a truly typological renovation, and generate an innovative and complex urban fabric which is therefore quite removed from those of conventional planning projects.

Overall, we see projects with varying origins, different organizational logics, and different temporal horizons, but which form part of the objectives which the city has pursued since 1992.

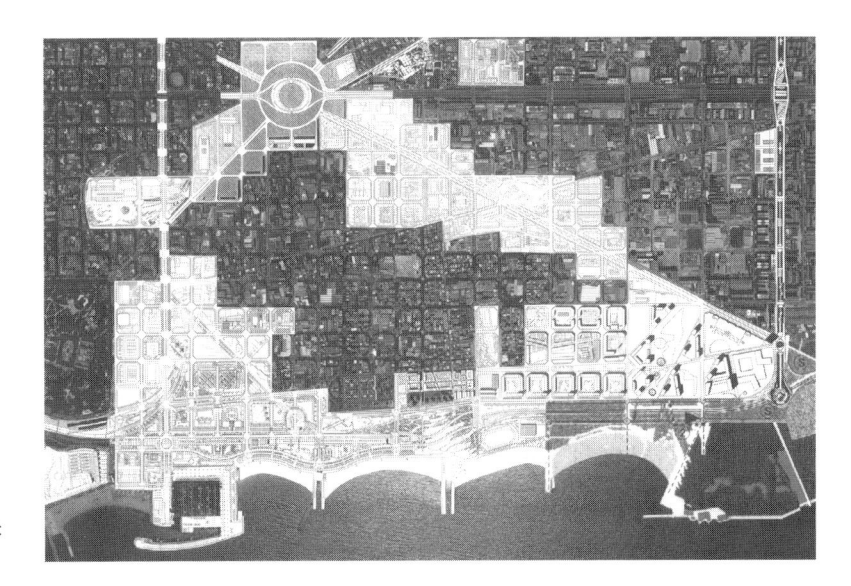

7.8
Aerial view of proposed development zones around Poblenou, late 1990s

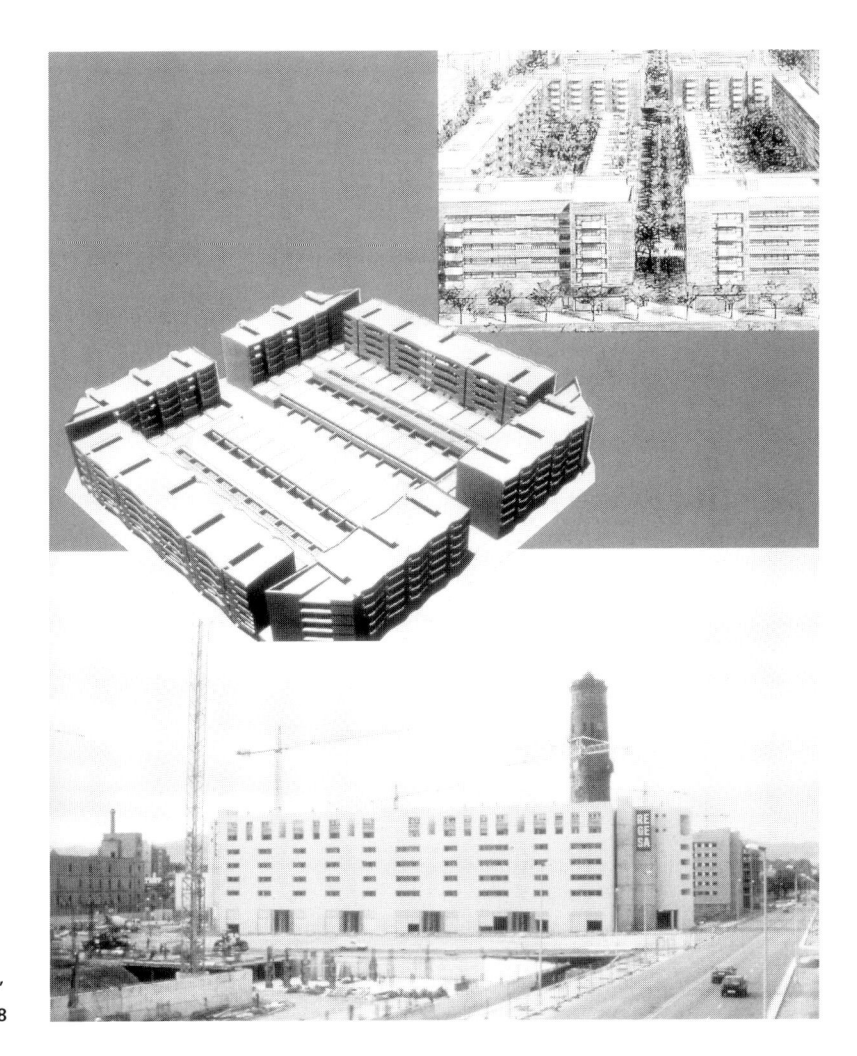

7.9
One of the blocks in the "five blocks" scheme in Poblenou, built about 1998

THE PLANNING PROJECT 145

There has been a high level of private participation in the processes of transformation, which in total make up more than 500 hectares of land recycling in the city, representing approximately 8% of the total urban area of the municipality. These projects propose the construction of more than 16,000 new dwellings, without counting those that are included in the reorientation of the current industrial sectors, of which at least 25% must be dwellings at an affordable price, as laid out in the specifications of the corresponding plans.

These are projects that seek to satisfy the ever-present desire to bring value to the periphery and improve the centre, and also to obtain from the space which Barcelona occupies all the benefit — social, economic and cultural — which corresponds to its position within the metropolitan region.

References

The first section contains a series of publications by the city council itself on planning matters. The series was begun in 1982 and constitutes a valuable resource for understanding the process of change and development undertaken.

Àrea d'urbanisme (1983) *Plans i projectes per a Barcelona, 1981–1982*, Barcelona: Ajuntament de Barcelona, 297 pp. (under the direction of Oriol Bohigas).

Les vies de Barcelona, materials del seminari de maig 1984, Barcelona: Ajuntament de Barcelona i Corporació Metropolitana de Barcelona (1984) 57 pp. (under the direction of Joan Busquets and José L. Gómez Ordoñez; texts in Catalan and Spanish).

Inicis de la urbanística municipal de Barcelona. Mostra dels fons municipals de plans i projectes d'urbanisme 1750–1930, Barcelona: Ajuntament de Barcelona i Corporació Metropolitana de Barcelona (1985) 277 pp. (under the direction of Manuel Torres Capell; catalogue of the exhibition at the Saló del Tinell, Barcelona, February and March 1985).

Àrea d'Urbanisme (Ajuntament de Barcelona) (1983) *Estudi de l'Eixample,* Barcelona: Regidoria d'Edicions i Publicacions (Ajuntament de Barcelona) 28 pp. (under the direction of Joan Busquets and the Laboratori d'Urbanisme de Barcelona; catalogue of the exhibition at the Casa Elizalde, Barcelona, 29 December 1983 to 28 February 1984).

Àrea d'Urbanisme i Obres Públiques and Àrea de Relacions Ciutadanes (Ajuntament de Barcelona) (1991) *Àrees de nova centralitat. New downtowns in Barcelona*, Barcelona: Ajuntament de Barcelona, 71 pp. (under the direction of Joan Busquets; text in Catalan, Spanish and English).

Àrea d'Urbanisme i Obres Públiques (Ajuntament de Barcelona); fundació Joan Miró (1987) B*arcelona, espais i escultures 1982–1986*, Barcelona: Regidoria d'Edicions i Publicacions (Materials de disseny urbà, 11), 167 pp. (under the direction of Lluís Hortet and Miquel Adrià).

Serveis de Planejament Urbanístic (Ajuntament De Barcelona) (1987) *Urbanisme a Barcelona. Plans cap al 92*, Barcelona: Àrea d'Urbanisme i Obres Públiques (Ajuntament de Barcelona), 194 pp. (under the direction of Joan Busquets).

Direcció de Serveis d'urbanisme (Corporació Metropolitana de Barcelona) (1987) *Projectar la ciutat metropolitana: obres, plans i projectes, 1981–1986*, Barcelona: Corporació Metropolitana de Barcelona, 255 pp. (under the direction of Juli Esteban).

Corporació Metropolitana de Barcelona (1987) *Pla de costes: proposta d'ordenació dela zona costanera metropolitana de Barcelona*, Barcelona: Corporació Metropolitana de Barcelona, 227 pp. (under the direction of Lluís Cantallops).

Patronat Metropolità del Parc de Collserola (1990) *Parc de Collserola. Pla especial d'ordenació i protecció del medi natural: realitzacions 1983–1989*, Barcelona: Mancomunitat de Municipis of the Àrea Metropolitana de Barcelona, 163 pp. (under the direction of Miquel Sodupe).

Regidoria de Programació i Pressupostos (Ajuntament de Barcelona) (1992) *Imatges del procés de transformació urbana, Barcelona 1987–1992*, Barcelona: Ajuntament de Barcelona, 35 pp. (coordinator: Artur Ferrer i Escriche; text in Catalan, Spanish and English).

Ajuntament de Barcelona (1993) *Barcelona, espai públic. Homenatge a Josep Maria SerraMartí*, Barcelona: Regidoria d'Edicions i Publicacions (Ajuntament de Barcelona), 207 pp. (under the direction of Rafael de Càceres and Montserrat Ferrer).

La rehabilitació de l'Eixample 1987–1991, Barcelona: Regidoria d'Edicions i Publicacions (Ajuntament de Barcelona) (1993) 159 pp. (collection from the exhibition at the Casa Elizalde, "La rehabilitació de l'Eixample en marxa 1987–1991", Barcelona, 6 June to 28 July 1991, (text in Catalan, Spanish and English)).

Ajuntament de Barcelona (1996) *Barcelona, la segona renovació*, Ajuntament de Barcelona, 222 pp.

Other references

Ajuntament de Badalona (1991) *Badalona amb el 2000 a l'horitzó*, Badalona: Àrea de l'Alcaldia of the Ajuntament de Badalona, 85 pp. (text in Catalan and Spanish).

Ajuntament de Barcelona (1996) *Barcelona, posa't guapa. Deu anys de campanya*, Barcelona: Ajuntament de Barcelona, 229 pp.

Ajuntament de Barcelona (1993) *Campanya per a la protecció i millora del paisatge urbà 1986–1992. Barcelona posa't guapa. Memòria d'una campanya*, Barcelona: Ajuntament de Barcelona, 231 pp.

Ajuntament de Barcelona (1984) *Pla General 1984–1992 i Programa d'Actuació Municipal 1984–1987 aprovat pel Consell Plenari de 18 de novembre 1983*, Barcelona: Gabinet Tècnic de Programació (Pressupostos, programes i memòries, 9), 47 pp.

Ajuntament de Barcelona (1988) *Programa d'Actuació Municipal 1988–1991 aprovat pel Consell Plenari de 4 de desembre de 1987*, Barcelona: Gabinet Tècnic de Programació (Pressupostos, programes i memòries, 12), 43 pp.

Ajuntament de Barcelona (1992) *Programa d'Actuació Municipal 1992–1995 aprovat inicialment pel Consell Plenari de 3 de març de 1992, aprovat definitivament pel Consell Plenari de 29 de maig de 1992*, Barcelona: Gabinet Tècnic de Programació (Pressupostos, programes i memòries, 18), 78 pp.

Ajuntament de Barcelona (1996) *Programa d'Actuació Municipal 1996–1999 aprovat inicialment pel Consell Plenari de 22 de desembre de 1995, aprovat definitivament pel Consell Plenari de 15 de marc de 1996*, Barcelona: Direcció de Serveis Editorials (Pressupostos, programes i memòries, 19), 89 pp.

Ajuntament de Sant Adrià de Besòs (1991) *Sant Adrià 12 anys de democràcia municipal*, Sant Adrià de Besòs: Ajuntament de Sant Adrià de Besòs, 124 pp. (text in Catalan and Spanish).

Ajuntament de Santa Coloma de Gramenet (1991) *Ciutat dibuixada: ciutat construïda* (s. l.), Ajuntament de Santa Coloma Gramenet, 63 pp.

Àrea d'urbanisme (Ajuntament de l'Hospitalet de Llobregat) (1987) *L'Hospitalet d'avui a demà: una proposta per ordenar i millorar la ciutat*, l'Hospitalet de Llobregat, Àrea d'Urbanisme (Ajuntament de l'Hospitalet de Llobregat), 137 pp.

Àmbit d'Urbanisme i Medi Ambient (Ajuntament de Barcelona) (1995) *Memòria 1991–1994*, Barcelona: Ajuntament de Barcelona, 180 pp.

Àmbit d'Urbanisme i Serveis Municipals (Ajuntament de Barcelona) (1992) *Memòria 1987–1991*, Barcelona: Ajuntament de Barcelona, 158 pp.

Arenas, M. *et al.* (1995) *Barcelona Transfer, Sant Andreu-La Sagrera, planificación urbana 1984–1994*, Barcelona: Actar, 160 pp. (text in Catalan, Spanish and English).

Barcelona New Projects (1994) Barcelona, (s. n), 127 pp. (catalogue of the exhibition at the Saló del Tinell, Barcelona, 29 March to 25 May 1994, text in Catalan, Spanish and English).

Barcelona Olímpica: la ciutat renovada (1992) Barcelona: Àmbit Serveis Editorials (HOLSA), 340 pp.

Bohigas, O. (1985) *Reconstrucció de Barcelona*, Barcelona: Edicions 62 (Llibres a l'abast, 198), 302 pp.

Busquets Grau, J. (1992) *Barcelona. Evolución urbanística de una capital compacta*, Madrid: Mapfre (Ciudades de Iberoamérica, 12), 425 pp.

Gabancho, P. and Freixa, F. (1995) *La conquesta del verd. Els parcs i jardins de Barcelona*, Barcelona: Regidoria d'Edicions i Publicacions (Ajuntament de Barcelona), 260 pp.

Guia la Barcelona del 93 (1990) Barcelona: Ajuntament de Barcelona, 130 pp. *L'impacte econòmic dels jocs olímpics de Barcelona 92* (1992) Barcelona: Regidoria de Programació i Pressupostos (Ajuntament de Barcelona), 35 pp. (working document).

Planes, L. (ed.) (1995) *Guia dels parcs de l'Àrea Metropolitana de Barcelona*, Barcelona: Àrea Metropolitana de Barcelona.

Primera Tinència d'Alcaldia (Ajuntament de Barcelona) (1987) *Memòria de la primera Tinència d'Alcaldia 1983–1987*, Barcelona: Ajuntament de Barcelona, 120 pp.

Secció de Disseny Urbà (Ajuntament de Santa Coloma De Gramenet) (1987) *Projectes: abril 1987*, Santa Coloma de Gramenet: Ajuntament de Santa Coloma Gramenet, 56 pp.

Servei d'Urbanisme i Habitatge (Ajuntament de Santa Coloma de Gramenet) (1987) *Perifèria o marginalitat: tres anys d'urbanisme municipal a Santa Coloma de Gramenet (1984–1987)*, Santa Coloma de Gramenet: Ajuntament de Santa Coloma Gramenet, 118 pp.

Various authors (1997) *Els 20 anys del Pla General Metropolità de Barcelona*, Barcelona: Institut d'Estudis Metropolitans de Barcelona (Papers Regió Metropolitana de Barcelona: territori, estratègies, planejament, 28), 105 pp.

Various authors (1991) *Barcelona: la ciutat central*, Barcelona: Institut d'Estudis Metropolitans de Barcelona (Papers Regió Metropolitana de Barcelona: territori, estratègies, planejament, 5), 52 pp.

Various authors "La ciutat davant el 2000", *Barcelona Metròpolis Mediterrània*, 15 (summer 1990), 65–128 pp. (Quadern Central).

Various authors "Collserola, el parc natural per descobrir", *Barcelona Metròpolis Mediterrània*, 14 (winter 1989–1990), 65–128 pp. (Quadern Central).

Various authors (1993) *La conurbació barcelonina: realitzacions i projectes*, Barcelona: Institut d'Estudis Metropolitans de Barcelona (Papers Regió Metropolitana de Barcelona: territori, estratègies, planejament, 13), 93 pp.

Various authors (1985) "La Rehabilitació de Ciutat Vella", *Barcelona Metròpolis Mediterrània*, 1, 103 pp. (Quadern Central).

Various authors "Circulació: conflicte i repte", *Barcelona Metròpolis Mediterrània*, 8 (summer 1988), 65–127 pp. (Quadern Central).

Various authors "La reconquesta del litoral Barceloní", *Barcelona Metròpolis Mediterrània*, 12 (summer 1989), 66–131 pp. (Quadern Central).

Various authors (1993) *Revitalització social, urbana i econòmica: documents*, Barcelona: Promoció de Ciutat Vella, SA, Districte Ciutat Vella (Ajuntament de Barcelona), 141 pp. (Second Ciutat Vella Conference, Barcelona, 2–4 December 1991).

Various authors (1991) *Revitalització urbana, econòmica i social: documents*, Barcelona: Promoció de Ciutat Vella, SA, Districte Ciutat Vella (Ajuntament de Barcelona), Barcelona més que mai, 189 pp. (First Ciutat Vella Conference, Barcelona, 6–10 November 1989).

Notes

1   An instrument included in planning legislation to order, in a precise way, the specific scope of urban land, in a similar manner to the partial plans in urbanizable land.

2   As per Bohigas, O. in Àrea d'Urbanisme (Ajuntament de Barcelona) (1983) *Plans i projectes per a Barcelona, 1981–1982*, Barcelona: Ajuntament de Barcelona, 297 pp. (under leadership of Oriol Bohigas).

3   As per Busquets, J. in Serveis de Planejament Urbanístic (Ajuntament de Barcelona) (1987) *Urbanisme a Barcelona. Plans cap al 92*, Barcelona: Ajuntament de Barcelona, 194 pp. (under leadership of Joan Busquets).

4　As per Busquets, J. in Serveis de Planejament Urbanístic (Ajuntament de Barcelona) (1987) *Urbanisme a Barcelona. Plans cap al 92*, Barcelona: Ajuntament de Barcelona, 194 pp. (under leadership of Joan Busquets).

5　Busquets, J. in Serveis de Planejament Urbanístic (Ajuntament de Barcelona) (1987) *Urbanisme a Barcelona. Plans cap al 92*, Barcelona: Ajuntament de Barcelona, 194 pp. (under leadership of Joan Busquets).

6　Área d'Urbanisme i Obres Públiques i Área de Relacions Ciutadanes (Ajuntament de Barcelona) (1991) *Àrees de nova centralitat. New downtowns in Barcelona*, Barcelona: Ajuntament de Barcelona, 71 pp. (under the leadership of Joan Busquets); Serveis de Planejament Urbanístic (Ajuntament de Barcelona) (1987) *Urbanisme a Barcelona. Plans cap al 92*, Barcelona: Ajuntament de Barcelona, 194 pp. (under leadership of Joan Busquets).

7　*Les vies de Barcelona. Materials del seminari de maig de 1984* (1984) Barcelona: Ajuntament de Barcelona i Corporació Metropolitana de Barcelona, 57 pp. (under the leadership of Joan Busquets and José L. Gómez Ordóñez).

8　*Les vies de Barcelona. Materials del seminari de maig de 1984* (1984) Barcelona: Ajuntament de Barcelona i Corporació Metropolitana de Barcelona, 57 pp. (under the leadership of Joan Busquets and José L. Gómez Ordóñez).

9　Direcció de Serveis d'Urbanisme (Corporació Metropolitana de Barcelona) (1987) *Projectar la ciutat metropolitana: obres, plans i projectes 1981–1986*, Barcelona: Corporació Metropolitana de Barcelona, 255 pp. (coordinated by Juli Esteban).

10　Direcció de Serveis d'Urbanisme (Corporació Metropolitana de Barcelona) (1987) *Projectar la ciutat metropolitana: obres, plans i projectes 1981–1986*, Barcelona: Corporació Metropolitana de Barcelona, 255 pp. (coordinated by Juli Esteban).

11　Direcció de Serveis d'Urbanisme (Corporació Metropolitana de Barcelona) (1987) *Projectar la ciutat metropolitana: obres, plans i projectes 1981–1986*, Barcelona: Corporació Metropolitana de Barcelona, 255 pp. (coordinated by Juli Esteban); Corporació Metropolitana de Barcelona (1987) *Pla de costes. Proposta d'ordenació de la zona costancra metropolitana de Barcelona*, Barcelona: Corporación Metropolitana de Barcelona, 227 pp. (coordinated by Lluís Cantallops); Patronat Metropolità del Parc de Collserola (1990) *Parc de Collserola. Pla especial d'ordenació i protecció del medi natural: realitzacions 1983–1989*, Barcelona: Mancomunidad de Municipios del Área Metropolitana de Barcelona, 163 pp. (under the leadership of Miquel Sodupe).

12　Corporació Metropolitana de Barcelona (1987) *Pla de costes. Proposta d'ordenació de la zona costanera metropolitana de Barcelona*, Barcelona: Corporación Metropolitana de Barcelona, 227 pp. (coordinated by Lluís Cantallops)

13　Patronat Metropolità del Parc de Collserola (1990) *Parc de Collserola. Pla especial d'ordenació i protecció del medi natural: realitzacions 1983–1989*, Barcelona: Mancomunidad de Municipios del Área Metropolitana de Barcelona, 163 pp. (under the leadership of Miquel Sodupe)

14　Àrea d'Urbanisme (Ajuntament de Barcelona) (1983) *Plans i projectes per a Barcelona, 1981–1982*, Barcelona: Ajuntament de Barcelona, 297 pp. (under leadership of Oriol Bohigas); Àrea d'Urbanisme i Obres Públiques (Ajuntament de Barcelona) and Fundació Joan Miró (1987) *Barcelona, espais i escultures 1982–1986*, Barcelona: Ajuntament de Barcelona (Materials de disseny urbà, 11), 167pp. (coordination by Lluís Hortet and Miquel Adrià, with Josep A. Acebillo as director of Projectes Urbans); Regidoria de Programació i Pressupostos (Ajuntament de Barcelona) (1992) *Imatges del procés de transformació urbana. Barcelona 1987–1992*, Barcelona: Ajuntament de Barcelona, 167pp. (coordination by Artur Ferrer i Escriche); Ajuntament de Barcelona (1993) *Barcelona, espai públic. Homenatge a Josep Maria Serra Martí*, Barcelona: Ajuntament de Barcelona, 167pp. (coordination by Rafael de Cáceres and Montserrat Ferrer).

15　Àrea d'Urbanisme (Ajuntament de Barcelona) (1983) *Estudi de l'Eixample*, Barcelona: Ajuntament de Barcelona, 167pp. (under the coordination of Joan Busquets, Laboratori d'Urbanisme de Barcelona. Catalogue of the exhibition in the Casa Elizalde, December 1983 to February 1984); Ajuntament de Barcelona (1993) *La rehabilitació de l'Eixample*

1987–1991, Barcelona: Ajuntament de Barcelona, 167pp. (under the direction of Amador Ferrer; the collection of the exhibition in Casa Elizalde, June–July 1991).
16 Ajuntament de Barcelona (1996) *Barcelona. La segona renovació*, Barcelona: Ajuntament de Barcelona, 167pp. (under the direction of Ricard Fayos).
17 Ajuntament de Barcelona (1996) *Barcelona. La segona renovació*, Barcelona: Ajuntament de Barcelona, 167pp. (under the direction of Ricard Fayos)
18 See Chapter 11 by Oriol Clos on the 22@ project.

Chapter 8

Public spaces in Barcelona 1980–2000

Nuria Benach

This chapter presents a vision of new public spaces in Barcelona in the past twenty years. It goes beyond stressing the transformation of the urban landscape and its contribution to the changing of the city's image. I have made some contributions in this direction (Benach 1993), thus adding myself, perhaps involuntarily, to the "architectural determinism" that has dominated the discourse on the city in the past decades. This architectural determinism, both in its positive and negative versions, has identified the analysis of what happened in the city with its spatial transformations, as if the social life of those spaces was to be construed automatically from them. This is not a new practice in urban studies but, nevertheless, it still makes it difficult to set up an adequate reading of urban spaces that captures the complexity of symbols, perceptions and uses that lived spaces involve (Lees 1994).

In an endeavour to set up a more comprehensive framework for analysing public spaces in Barcelona, I will trace three different ways of looking at the same issue, to show that a real understanding of public space needs a combination of all these elements. The three visions that follow will bring to mind the spatial framework proposed by Henri Lefebvre (1974) in his "three moments of social space". For Lefebvre, space could be perceived, conceived or lived, giving an interpretation capable of conveying these three interrelated visions of space. Although the idea developed in this chapter is inspired by this theoretical proposal (and even more by the passionate reading of this work by Edward Soja in 1996), I will make use of it only to encourage an analysis of Barcelona urban public spaces that overcomes architectural determinism and expands our understanding in other directions.

According to Lefebvre's framework, *perceived space* is "the materialized, socially produced, empirical space, directly experienced, open, within limits, to accurate measurement and description. It is the traditional focus of attention in all spatial disciplines ..." (Soja 1996, p. 66). In relation to new urban public spaces in Barcelona, it is about the physical spaces that architects have built; it is the spaces of stone, the *spaces of design*, the spaces that citizens observed and assessed before occupying them with different motivations and results.

Conceived space is a "conceptualized space, the space of scientists, planners, urbanists, technocratic subdividers all of whom identify what is lived and what is perceived with what is conceived" (Soja 1996, pp. 66–67). This is the dominant space, the space of power, of regulation. When analysing public space in Barcelona, this will be the *space of renovation*, with a special reference to those public spaces that were conceived and built to renew their immediate urban environment, thus contributing to the renewal of the urban tissue. Not least, these public spaces were also used in a very remarkable way as symbols of the "revitalized city" (Benach 1993).

Lived space, finally, refers to social space in all its complexity, filled with complex symbolisms. It is the space as appropriated by its inhabitants and users with the capacity to generate qualitatively different spaces, virtual "counterspaces", spaces of resistance to the dominant order (Soja 1996, p. 68). In Barcelona, public space is commonly analysed as a conceived or a perceived space much more than as a lived space. This means putting people back in the public spaces, as a necessary condition of understanding and assessing the spaces in all their complexity. It will then be the *space of people*, with different uses. These uses will often be distinct from those for which the spaces were designed, spaces of contact and of conflict, spaces of socialization, of entertainment, of fear.

First: spaces of design

From the very beginning of the 20-year process of transformation of Barcelona, new public spaces played an outstanding part. At first, they were a central characteristic of the urban processes of renovation that culminated in the organization of the Olympic Games in 1992. After that, they were part of a distinctive way of conceiving and designing a competitive city ("a model", as described by its promoters). The chronology shows how fast and effective the process was. In 1979, the first democratic municipal elections were celebrated. In 1980, the architect Oriol Bohigas arrived in the city council. And, in 1983, the inauguration of public spaces started. There were new public spaces such as the plaça Sóller or the plaça del Països Catalans, reformed spaces such as the Plaça Reial or the plaça de la Mercé, and the placing of sculptures in public spaces, such as the monument to Picasso by Tàpies, the piece *Dona i Ocell* by Joan Miró in the then unfinished plaça de l'Escorxador, or the structure *Mediterrània* by Xavier Corberó in the plaça Sóller. Thus, in less than four years, the basis for the creation of public spaces was not only well established ("monumentalizing" the periphery, opening out the historic centre, setting up sculptures by prestigious artists), but the first results emerged.

8.1
New open space in Gràcia, 1980s

The rhythm with which these projects were carried out, especially during these first years, was very intense (afterwards the Olympic goal tended to monopolize investments). The most important outcomes in the period 1979–82 were the renewal of the plaça Reial, the passeig de Picasso, the moll de la Fusta, the squares of the Gràcia district or the plaça Sóller (Ajuntament de Barcelona 1983, p. 97) (Figure 8.1), and, in the next four-year period, the parc de l'Escorxador (1985), the parc de l'Espanya Industrial (1985), the parc de la Pegaso (1986) and the parc de la Creueta del Coll. In the following years, the creation of medium-sized new parks and small urban gardens was continued, and the new symbolic places of the centre of the city such as the axis Rambla-plaça de Catalunya or the plaça de la Catedral, were also the objects of intervention (Ajuntament de Barcelona, 1992). In these years, the remodelling of the sea front started, while new, bigger urban public spaces such as the parc de l'Estació del Nord or the parc de Renfe-Meridiana were designed. In the 1990s, the public spaces policy continued.

In quantitative terms, during the decade 1982–92, more than 200 hectares of park had been gained, largely thanks to the reconversion of former industrial spaces (as Fabré and Huertas, 1989, have shown), while during the 40 years of the Franco period only 70 hectares were created (Cáceres and Ferrer 1993). The peripheral location of many of these public spaces is to be explained by the existence of large sites in those districts. The intervention in the centre and in some historical areas such as Gracia or Sants, although significant, consisted of small operations.

One outcome of the design of new spaces was the improvement of many neighbourhoods, but even more important was the creation of prestigious spaces, thanks to their architectural design and the use of street sculptures.

Lluís Permanyer, who has often acted as a kind of unofficial chronicler of the city, has remarked on the contribution of these art works not only to the monumentalization of the urban periphery, but also to the competitiveness of the city:

> By means of such a policy, not only the urban landscape has been enriched, but it has also been capable of giving identity at strategic points to neighbourhoods lacking history and personality. Thus, it is neither strange nor surprising that one of the most degraded areas, Nou Barris, has been one of the districts which has gained most from this policy. This open air museum contrasts with the low quality of the sculptures created recently in Madrid, as well as with those in Paris dedicated to the commemoration of the bicentennial of the Revolution (Permanyer 1996).

The architect Ignasi de Solà-Morales maintained that sculptures were regarded as architectural elements "not to contribute to a museographic idea, but to their use in urban space" (Solà-Morales 1987). The most classical, although disputed, examples are the plaça de Sóller (sculpture by Xavier Corberó) and the plaça de la Palmera (with two concrete walls by Richard Serra). In other cases, however, the artistic contribution was no more than an added prestigious element in a public space that already existed. The function of the sculptures as landmarks (in the Lynch, 1960, sense) is without doubt: they are used as a focus of attention, where eyes previously passed by, they mark the entrance to the new symbolic spaces (for instance, Roy Lichenstein's Barcelona head at the entrance of the Old Harbour, or Frank Gehry's golden fish close to the two Olympic towers on the sea front).

Second: spaces of renovation

Fifteen years of creation of new public spaces have generated some thought about the goals of policy, and about the means used to achieve them. Public spaces were said to be aimed, first of all, at solving the historical deficiencies of the city, a city made too dense by the uncontrolled and speculative urban activity of the predemocratic period.[1] The political pressures of the neighbourhood associations, which had been very active at the end of the 1970s, then had to face a democratic city council that seemed to share some of their aspirations. J. M. Abad, in charge of the urban planning department in that first democratic council, and the person with prime responsibility for the organizing committee of the Olympic Games of 1992, explained the goals of planning policy:

> A basic goal will be the achievement of a balanced Barcelona, eliminating segregation and searching for social and territorial

equity for all citizens, in the access to social facilities and in the urban quality of their neighbourhoods, preventing the expulsion of the lower income groups from the city centre. Our goal is to improve facilities and to remodel the most degraded neighbourhoods by the initiative of the public sector; we aim to put in practice a policy that allows for the upgrading and rehabilitation of the old cores and prevents the current process of accelerated degradation and higher densities, while trying to avoid as much as possible the speculative retention of land and unoccupied dwellings (Ajuntament de Barcelona 1983).

The new public spaces, however, were designed not only to settle historic debts, but to influence actively the urban environment. Oriol Bohigas, the key figure in sketching the major lines of the Barcelona urban planning during the 1980s, has been repeatedly very explicit on this:

Among the methods and the specific instruments of the Urban Planning Department we have a clear direction. We will proceed directly with public spaces with two goals: to make space of quality and at the same time to create a focus that can generate spontaneous transformations. It is evident that when a public space is built or rebuilt in a degraded neighbourhood, this is a focused intervention, the motor of the regeneration of the environment, stimulated by the users themselves (Bohigas 1985, p. 21).

The same idea, in even more explicit terms, was expressed by Bohigas, when he affirmed that this urban intervention was "metastatic, strategic, for reconstruction, and mainly supported by the design of public spaces, since they are the most immediately effective to achieve these goals" (Bohigas 1987, p. 12). They are metastatic, because "a series of actions can be the focus of regeneration for the surroundings"; strategic, since "in order that metastasis can be effective, the initial 'infection' has to be applied to the nerve centre of the neighbourhood, of the city, of the metropolis", and for reconstruction, because "to build in what is already built, to improve what already exists, to transform, to modify, to rehabilitate, to resignify, to underline or to create identities are the clearest and most important objectives".

A political stance was also present. Pasqual Maragall stressed the goal of social justice that was inherent in the policy, although this was, essentially, the same as the one formulated by the architect:

We have applied the best design to the working class districts, with more emphasis and passion than in the traditional districts, because the goal is that these districts have to be the new city, not only chronologically but also in terms of their tone and quality. This newer Barcelona has to be the most emblematic exponent of the new Barcelona as a whole (Maragall 1991a).

It is easy to demonstrate that a very substantial part of public investments in public spaces has been made in those peripheral working class districts. What is not so clear, however, is that those who have gained most from these transformations have been those who live in these districts. Rosa Tello, analysing public investments up to 1994, observed that the likely improvement of these districts as a consequence of their monumentalization was followed by a rise in house prices above the city average (Tello 1997). Furthermore, it remains to be proved that these spaces were the appropriate ones for the subsequent social use that they have had.

In Ciutat Vella, the historical city, the reasoning used is very similar, even though the context is rather different. Very high density has led to the strategy of opening out the urban texture, using small pieces of land to get new open spaces. The monumentalization used in the peripheral working class districts was replaced, in this case, by a reinterpretation of the old buildings and historical spaces. Despite the enormous attractiveness that the old city potentially has, the social tension that a massive intervention would have provoked, has meant that all action has been cautious. The Raval neighbourhood (traditionally known as the Barrio Chino or "chinatown") had a very strong stereotype as a site of drug addiction, prostitution and criminality. For this reason, its proposed transformation from a degraded district to a cultural centre (attractive enough for new high-income inhabitants, who would never choose the monumentalized periphery to live in) *had to seem* an irreversible process for it to be implemented. In 1987, when interventions in the Raval were still more on paper than in reality, Pasqual Maragall felt very confident about it:

> It is like an embroidery project, very slow, that has to be done very carefully, that will demand a big effort, *but a project that is there, that is drawn, that can be done, and that will be done*. So, the Raval, now a focus of nostalgia, if you like, and of problems, will be converted step by step into a centre of culture and of cultural activity (Maragall 1987, pp. 90–91, emphasis added).

Fourteen years later, the process of renovation of the Raval is already visible, although the process of gentrification is still not very clear (Martínez 2000). Many new small spaces have been created through small-scale operations, alongside a few bigger and deeper interventions, such as the plaça dels Angels, just in front of the Museum of Modern Art designed by the architect Richard Meier, or the new avenue called Rambla del Raval, a paradigmatic example of a new space designed to influence its surroundings more than to have a sense in itself (Figure 8.2).

Third: spaces of people

Public space is always a response to a particular conception of urban space. Historically, it has been a key element of urban intervention and very often

8.2
Rambla del Raval, completed in 2001

8.3
Public space in front of Museum of Contemporary Art of Barcelona (MACBA)

it has had an added symbolic power that has reinforced or legitimated urban plans. But public space is also space used by citizens, space that is adaptable and has a changing meaning. As against private space (space of safety, space of order), public space is always a space of uncertainty, of entertainment, of conflict, where new situations and possibilities can be generated (Wilson 1995).[2] Public space is also the space of socialization, the space of cultural contact, the space of collective action. For this reason, public space should be considered not only for its design or in terms of the idea behind plans; the ways in which it is interpreted, used and appropriated by its users should also be analysed. It is not only through its design that a public space has civic meaning; its identity and its symbolic power are acquired only by means of social action, often with high social costs (Lees 1994). In this sense, it should be noted that public spaces and civic spaces are not necessarily the same thing; civic spaces are those that have been used as vehicles for social action, spaces with "real social significance" even though they have not been accepted as such (Hayden 1995).

In Barcelona, we have been used to contemplate public spaces only as those that were conceived to accomplish specific planning goals. However, there are many other public spaces, some with little social content, others with a very high significance, but completely different from the intended meaning (Figure 8.3). Public space in Barcelona has been taken over by those who have something to say (as the recent demonstrations in favour of the anti-globalization movement have shown), by those who have no other place to spend their leisure time (as the proliferation of immigrants in streets, squares and open spaces has shown), by those who have no private space to live in (as shown by the appropriation of symbolic spaces such as plaça Catalunya by homeless immigrants without official papers).

A clear disjunction between spaces as they were designed and the social use they finally have can often be seen; sometimes space is the vehicle that expresses social tensions. In the Raval, for example, the

8.4
Public space in Ciutat Vella, built in the 1980s

existence of social conflicts has been perceptible in the past few years. The press has echoed recent violent confrontations among local and immigrant people,[3] and the City Council has already initiated schemes to preserve the safety of public space. Several years ago, it was decided that some squares would be closed at night (Figure 8.4) and in 2000 the first video camera was installed in a new public space, by chance the one that is named after George Orwell. Public space can be the space of socialization, but it can also be a space of conflict and fear.

Ways of thinking about public spaces in Barcelona

"City" and "people" are two words that have been very commonly used together in discourses about the recent transformations of Barcelona. In 1987, Pasqual Maragall proclaimed that "the city is the people", apparently meaning that the main goal of Barcelona's policies during the following years was going to be for people. But since then people have been regarded as passive spectators more than as the main protagonists of changes. At the moments of urban life created most for spectacle, such as the Olympic period, people have been regarded above all as one more attractive feature of a lively city. They have been left as part of the stage set more than as actors of the play. When popular acceptance (lived space) of the new spaces has been stressed (and politicians seem never to have missed the opportunity for such emphasis),[4] this was taken as a sign of a positive judgement on new spaces from a material perspective (perceived space), thus legitimating the whole project of urban "revitalization" (conceived space). Blurring together (but not integrating) the perceived, the conceived and the lived has been a way to generate a hegemonic interpretation of urban transformations; an interpretation that seemed, especially thanks to its repetition, the only possible one. However, we should avoid this trap and, instead of setting the three perspectives against each other, we should look for a way to integrate them. Edward Soja's Thirdspace can be useful here. He uses the word Firstspace to designate the materiality of spatial forms and Secondspace for the ideas about space. The mixture of real and imagined places is what he calls Thirdspace. What Soja says, following Lefebvre, is that we should not concentrate on just one of these modes of thinking, but look for "the creation of another mode of thinking about space that draws upon the material and mental spaces of the traditional dualism but extends well beyond them in scope, substance and meaning." (Soja 1996, p. 11).

The city is the people, we have been repeatedly told. I would add that people should be also the city, so I would like to end this chapter by proposing a conceptualization of public spaces that takes into account the diverse forms of social appropriation that transform the city and give it authentic meaning. Precisely because hegemonic discourses on the city are not contested any longer with alternative discourses (which some people have interpreted as an absence of different views), but are contested through everyday practices, public spaces are the privileged site for resistance and for the generation of new possibilities. Public spaces, as spaces of contact[5] between different perceptions, uses, cultures and aspirations, can be read, in this way, as places where "all places are"[6] and, therefore, places where we could think about space with the new way of thinking that Soja has been demanding. We should not continue interpreting and judging Barcelona urban public spaces without such a new perpective.

References

Ajuntament de Barcelona (1987) *Memòria de la primera tinència d'alcaldia, 1983–1987*, Barcelona: Ajuntament de Barcelona.

Ajuntament de Barcelona. Ambit d'Urbanisme i Medi Ambient (1995) *Memòria 1991–1994*, Barcelona: Ajuntament de Barcelona.

Ajuntament de Barcelona. Ambit d'Urbanisme i Serveis Municipals (1992) *Memòria 1987–1991*, Barcelona: Ajuntament de Barcelona.

Ajuntament de Barcelona. Tinença d'Alcaldia de Planificacio i Ordenacio de la Ciutat (1983) *Memòria d'activitats 1979–1982*, Barcelona: Ajuntament de Barcelona.

Benach, N. (1993) "Producción de imagen en la Barcelona del 92", *Estudios Geográficos*, (LIV)212, pp. 483–505.

Bohigas, O. (1985) *Reconstrucció de Barcelona*, Barcelona: Edicions 62.

Bohigas, O. (1987) "Metàstasi i estratègia" en *Barcelona espai i escultures (1982–1986)*, Barcelona: Ajuntament de Barcelona, pp. 11–12.

Cáceres, R. and Ferrer, M. (eds) (1993) *Barcelona espacio público*, Barcelona: Ajuntament de Barcelona.

CAU (1974) *La Barcelona de Porcioles*, Barcelona: Laia.

Fabre, J. and Huertas, J. M. (1989) *La construcció d'una ciutat*, Barcelona: Plaza & Janes.

Hayden, D. (1995) *The Power of Place. Urband Landscapes as Public History*, Cambridge, Massachussets: MIT Press.

Lees, L. H. (1994) "Urban public space and imagined communities in the 1980s and 1990s", *Journal of Urban History*, (20)4, pp. 443–65.
Lefebvre, H. (1974) *La production de l'espace*, Paris: Anthropos.
Lynch K, (1960) *The Image of the City*, Cambridge, Massachusetts: MIT Press.
Maragall, P. (1987) *Per Barcelona*, Barcelona: Edicions 62.
Maragall, P. (1991a) *Barcelona, la ciutat retrobada*, Barcelona: Edicions 62.
Maragall, P. (1991b) *L'estat de la ciutat 1983–1990. Discursos de balanç d'any de Pasqual Maragall i Mira, Alcalde de Barcelona*, Barcelona: Ajuntament de Barcelona.
Marti, F. and Moreno, E. (1974) *Barcelona ¿a dónde vas?*, Barcelona: DIROSA.
Martínez, S. (2000) *El retorn al centre de la ciutat. La reestructuració del Raval entre la renovació i la gentrificació*, doctoral thesis, Universitat de Barcelona.
Permanyer, L. (1996) "Un museu d'escultures a l'aire lliure", *Barcelona, metròpolis mediterrània*, No. 29.
Pratt, M. L. (1992) *Imperial Eyes: Travel Writing and Transculturation*, New York: Routledge.
Soja, E. (1996) *Thirdspace*, Cambridge, Massachussets: Blackwell.
Solà-Morales, I. (1987) "Qüestions d'estil" in *Barcelona espai i escultures (1982–1986)*, Barcelona: Ajuntament de Barcelona, pp. 13–18.
Tello, R. (1997) "Les conseqüències de l'urbanisme dels darrers quinze anys", in Roca, J. (ed.) *El municipi de Barcelona i els combats pel govern de la ciutat*, Barcelona: Institut Municipal de Barcelona/Proa, pp. 277–84.
Wilson, E. (1995) "The Rhetoric of Urban Space", *New Left Review*, 209, pp. 146–60.

Notes

1. Some of the books that reflected the fighting spirit of the moment were published, significantly, in 1973 (CAU) and 1974 (Marti and Moreno).
2. The film of Pedro Almodóvar *All about my mother*, much of it filmed in Barcelona, suggested this contrast between the safety and predictability of private spaces as against the conflict and disorder of degraded and not yet urbanized public spaces. These latter were seen as giving the possibility for intense living, even though this might be hard.
3. For example, the hard words that an old inhabitant of the Raval confessed to a local newspaper about the Pakistani immigrants: "They go out everyday like beetles, you could not say how many of them are there behind each door" (*El Periódico* 02.08.2001).
4. "Every time that a new park, a new pedestrian street or a square has been opened, we have seen the spontaneous and lovely phenomenon of its immediate occupation by people" (Maragall 1991a, p. 51), or "Just after a new space is inaugurated in Barcelona, it is already full of people" (Maragall 1991b, p. 79).
5. The concept of *contact zone* developed by M.L. Pratt is very suggestive here. Although initially referring to spaces of colonial encounters, it can be very useful in the analysis of urban spaces (Pratt, 1992).
6. This is Borges's *Aleph* that Soja used to evoke new ways of looking at contemporary Los Angeles.

Chapter 9

Public space development in Barcelona — some examples

Jordi Borja, Zaida Muxí, Carme Ribas and Joan Subirats, Jaume Barnada, and Joan Busquets

Introduction (*Jordi Borja*)

In Barcelona, the idea of taking public space as the basis for "building the city" was, without doubt, the principal characteristic in democratic town planning in the 1980s, one that continued to predominate into the 1990s with the requirement for accessible, quality public spaces included in the great projects towards 1992.

The strategy of public spaces at all possible scales, from the "mini square" to parks, was based on politicians' and experts' confidence in the positive impact such action has on the environment. It was also, however, a pragmatic response to social demands, reinforced by decentralisation and the use of existing planning to reclaim land for public spaces and services, optimizing employment of the few financial resources available.

The cases we present illustrate the diversity of actions that exemplifies a strategic policy for public spaces:

- opening up of squares to promote the regeneration of degraded old centres (Pla del Raval);
- appropriation of empty spaces or road infrastructure works to create high quality avenues in the periphery (Via Júlia, Rambla de Prim, Ronda del Mig);

- rehabilitation of the inner courtyards of city blocks as public spaces (Eixample district);
- creation of public spaces through private commercial operations (Maremagnum);
- creation of city parks and walks by converting port and railway zones and rehabilitating obsolete facilities (sea front, Estació del Nord, Joan Miró/Escorxador Park).

Private public space: Maremagnum (*Zaida Muxí*)

New spaces for consumer activities combine leisure and entertainment facilities with shops. In most cases, the space takes the form of containers with no reference to their actual location. They are bubbles with their whole internal world, which recreates or is reminiscent of the city. Here we find constant nominal references to civic public spaces: such-and-such a square, such-and-such a street, etc., revealing a deliberate ambiguity as to whether these should be considered public or private spaces.

This new phenomenon of private spaces pretending to be public leads us to wonder whether the appearance of new consumer spaces, for the purchase of goods, entertainment and leisure, will in the future become socially defined, enriched, lived-in spaces as are today's squares, streets and parks, etc., without us questioning why or how they appeared.

The main difficulty for such acceptance concerns control and freedom. New consumer spaces are governed by rules and laws based on private property, on the right of admission; that is, on exclusion. This state of affairs, added to the time limits not found in public space, is in direct conflict with the civil rights associated with freedom of association, expression and circulation.

In any case, we may well extend our definition of public space, although this does not mean excluding or replacing one type of space with another. This should follow the path indicated by Marco Cenzatti and Margaret Crawford[1] towards a more flexible definition of public space in a continuously changing reality, in this way "interpreting these new spaces as the non-disappearance of public space and the emergence of a new type of public realm …"

In most cases, the containers behave blindly towards their immediate environment, being conceived as a simulation, their own fantasy world that denies belonging to or relating to the place, which they consider aggressive, unpleasant, uncontrollable. In some cases, the appearance of these "leisure and consumer containers" has helped to create a public space that indicates that an urban policy does exist, an idea that the city can be made up of both public and private parts.

Maremagnum is a leisure and shopping centre on Barcelona's Moll d'Espanya, in the old port, with multi-screen cinema, aquarium and Imax

9.1
Maremagnum and marina, Port Vell

cinema (Figure 9.1 and colour plate 6). The complex is an example which demonstrates that market logic is not the only logic that exists and that a centre of this type can be inserted into a planned city, taking the relation with the environment into account, responding to certain characteristics of the location and forming part of an urban whole without making the investment less profitable — quite the contrary, in fact.

The relation the project established with the Rambla helps this central urban boulevard to extend towards the port in the form of bridges. This achieves a two-fold goal: it helps to attract the large numbers of pedestrians that walk along the Rambla every day for commercial purposes, but also prolongs this space for the city.

In this way, a new visual and spatial relation with Barcelona is obtained, improving the quality as a public space of the part of the Rambla nearest the sea. This improvement is achieved not through coercion or imposition, but through the fact that the larger number of people here ensures diversity and, thereby, security.

It can be inferred from these results that security is not synonymous with exclusion but with diversification, and that acceptance of new consumer habits need not always imply negative results for the urban space.

The Rambla del Raval — an opportunity? (*Carme Ribas and Joan Subirats*)

For many, many years, there had been talk in Barcelona of the need to undertake determined, broad-ranging action to improve living conditions in the "Old City", or "Ciutat Vella" district, as it is known. This area comprises both the original core of the city and the zone occupied by an initial expansive phase, still inside the city walls: the Raval district. The origins of so-called "esponjament" or "opening-up" of the district go back as far as the times of

Ildefons Cerdà, Baixeres and the early twentieth-century "hygienists", though it was the GATCPAC projects (with Le Corbusier and without) that finally became best known. With the restoration of democracy, unaffected by the suspicion that earlier attempts during the Franco dictatorship had aroused, the city council launched an ambitious plan to redevelop and rehabilitate the neighbourhood. The overall balance of these years of massive intervention in Ciutat Vella is not negative. On the contrary, reasonably good work has been done. One in five houses in the neighbourhood has been rehabilitated, its former inhabitants rehoused. Over 21,000 million pesetas have been spent on public redevelopment, direct rehabilitation and public subsidies for private rehabilitation. And the truth is that, thanks to this and the arrival of new waves of emigrants, there are children and young people in the district again. This success is reflected in the many applications for new certificates of habitability declaring that dwellings are fit to live in.

Despite the many successes, however, the current situation obliges a number of questions to be raised. In our view, the complexity of the district's urban and human fabric was not taken into account in certain weighty interventions. Significant spaces have been *sventrato* (gutted) when they could easily have been rehabilitated. There has often been too much red tape applied in regulating new buildings. The architectural quality of new buildings is often inferior to what had been demolished. New streets have been built from zero without the least regard for local signs of identity (carrer Maria Aurelia Capmany). Squares have been built whose most notable characteristic is the sense of unease they arouse (Plaça Caramelles). Finally, an enormous space, the Rambla del Raval, with dimensions alien to the traditions of the district, has been opened up, its future still to be decided.

Other interventions have been made with more humility, care and *finezza*, opening up spaces that already seem to have been there for years, without causing scars that will be difficult to heal (Plaça de la Mercè, or the space of Allada i Vermell, which opens up after the Princesa–Assaonadors junction). In what should be seen, generally speaking, in a positive light, the Ciutat Vella experience shows us that there are some city zones that cannot be dealt with as just another neighbourhood. You cannot just come along, standard manual in hand, and demolish, mark out, establish rules and regulations, build and house residents in spaces that have become so complex over long years of history. There are problems of light, of density, of working with projects that speak the same language as that not demolished, which seek to rehabilitate without false conservationism but with respect and quality. It would have been better to work more "as a continuation of" than "as a replacement of".

Some of the old and new residents in this great empty receptacle that is today the Rambla del Raval look with eerie concern upon this great sun-filled space (Figure 9.2). In a district known for its back streets, damp and insalubrity there can never be too much sun, but there does exist a sensation of space out of all proportion, of frontier or no-man's-land, rather than of a

9.2
Rambla del Raval, showing trees well established 2003

shared square, helping to bind the district together. The Rambla del Raval and its future both pose a huge problem and present a great opportunity. It will be a problem if this space is not filled with urban and civic fabrics enabling progress to be made without diminishing diversity and cohesion. We all know that there is no such thing as a social vacuum. If the public powers cannot involve the residents, shopkeepers, associations, in governance, in sharing responsibility for this public space, other, clandestine, criminal fabrics will fill it. The public authorities should not monopolize, but build bridges, facilitate social self-government because, in spite of everything, the truth is that the Raval district is seeking its own personality behind these wounds, wounds that do not heal easily.

The multi-coloured vision of children leaving the local schools every day whisper to us that this district is a forerunner of what Barcelona is destined to become more and more. Through their work, district associations and municipal services are helping to maintain cohesion despite many attempts to fill vacuums with criminal fabrics. Streets such as Riera Alta, Riera Baixa, Carme and Hospital are becoming lined with much more interesting new and second-hand clothes shops than the predictable, repetitive chain stores found elsewhere. It is a delight to escape from McWorld shopping centres which are becoming ever less original, to stroll around these streets, which still conserve much of their traditional personality. Maghrebi butchers shops and Pakistani restaurants flank bars full of Erasmus grant students and artists seeking their big chance. The loss of culture and identity that affects cities all over the developed world has not yet penetrated Raval. No one wants a new Marais in Ciutat Vella. No one wants a lifeless yuppified district. What is at stake in Ciutat Vella is saving its *mestizaje*, its diversity, its pluralism of people and uses, and to do so with innovative respect for the identity of a neighbourhood that was Barcelona before Barcelona became what it is now.

In the final outcome, we want places that conserve their cultural diversity and decent quality of life. The rehabilitation of Ciutat Vella aims to achieve reasonable living standards, but we should all make sure that this is not done at the cost of losing identity, losing diversity of people and uses. We need density, we need complexity, we need people with simply complicated architectural ideas, not engineers with complicatedly simple ideas. Let local history and the present situation act as a positive constraint, let us not seek an impossible fresh start. Today's Rambla del Raval is an opportunity.

The new ramblas (*Jaume Barnada*)

Barcelona's urban history has produced a city in which public space is a scarce good, one which is often over-used. Barcelona is a city of streets. Streets have a fundamental importance that goes beyond the function of connecting: they are complex places of civil relations.

There is a huge diversity of streets in the city. For example, there are the streets of the Eixample district, where pedestrians have equal space with vehicles. Here, the long, tree-lined streets respond to the model of the modern city. Or the Rambla, without doubt one of the key centres of activity in the city and the urban public space *par excellence*. Barcelona is a place for walking.

In the early 1980s, a process of monumentalizing the periphery began in an attempt to regenerate and consolidate the city and to give structure to the new districts, urbanizing and creating urban itineraries. Via Júlia was one of the first examples of this democratic policy, and an extrapolation of the idea of the rambla in a new location. The initiative produced an unquestionable improvement and helped to initiate a process of economic and social regeneration in the area. The original space was little-used, empty land at one end of the city, a barrier to possible connections between the Roquetes and Prosperitat districts. The arrival of the metro linked these districts with the centre, but the really important action was the urban development project for Via Júlia that provided the necessary qualities of the city.

This and other urban development projects have brought about improvements in many spaces, both in the city centre and in the periphery: Rambla de Prim (colour plate 7), Rambla Catalunya, Passeig Lluís Companys, Rambla de Sant Andreu, etc. by making the thoroughfare a mixed space balancing traditional uses and modern civic activities.

This aim of improving the environment and facilitating complex urban uses within the public highway can be seen particularly in the recent covering of the Ronda del Mig ring road in the Sants and Les Corts districts. This maintains traffic flows whilst creating a space for pedestrians in the form of a rambla or boulevard along the covered "roof" of the fast road below. Such action prevents the urban fabric from being split up whilst establishing a new, diverse type of thoroughfare in which a variety of activities are made compatible, one with the other.

Barcelona's Eixample district — inner courtyards (*Joan Busquets*)

The importance of the Eixample district to the shape of central Barcelona is generally linked to that of its streets and boulevards and the continuity of its house fronts, made up of strikingly different architectures and uses. This urban fabric, built up over time according to the lines laid down in Cerdá's plan, has a complementary space: the inner courtyards of housing blocks. It would be difficult indeed to understand the force of the Eixample, its rigorous composition of house fronts, its variety of functions, without the space inside blocks, intended in the original plans not to be built on and onto which the inner face of houses look, a place of quietude and a domestic, interior image (Figure 9.3).

Although there are service or storage buildings in most inner courtyards, reclaiming them as green spaces can be of enormous benefit to residential and service structures. For this reason, after the completion of the Eixample Study,[2] Barcelona City Council proceeded to draft the Eixample Ordinance, approved in 1986. This stipulates that the transformation of all parks and blocks must enhance the central space in the heart of the block,

9.3
View of typical interiors of Eixample blocks, filled with buildings

even if use is private, by planting trees. This establishes a mechanism for opening up these internal spaces, an expanding process every stage of which improves on the last. Thanks to this process, green is more and more often seen in these courtyards or patios.

Moreover, around 30 of the 600 blocks in the district are gardened as one unit and allow public access to the interior: blocks such as Las Aigües, La Sedeta, etc. (Figure 9.4). The aim of this strategy is to ensure that there is an internal public space within a radius of five or six blocks. The strategic possibilities and the opportunities offered by each different block must be taken into account in rehabilitating this existing fabric (Figure 9.5).

In any case, the two strategies operate at different rhythms, on different timescales. Nonetheless they both seek to revive one of the lost dimensions in the great project of contemporary Barcelona.

9.4
New patio with garden and play area on Gran Vía (Bailén/Girona)

9.5
Patio of the Jardins d'Antoni Puigvert

New parks, reclaimed spaces (*Jaume Barnada*)

Barcelona needs internal green spaces. The Ciudadela Park is an example which is almost impossible to repeat in the consolidated city. Though unevenly urbanized, Montjuïc and Els Turons ("the hills") are rather inaccessible and atypical parks. Collserola is Barcelona's great green lung, but lies outside the city and is still little used by citizens. A future challenge is without doubt that of moulding this space into a real, central park within the metropolitan region.

In the late 1970s, there were in the city a number of obsolete factories on abandoned industrial sites, old buildings without heritage value but testimonies to history, in the Eixample district or in the first ring of traditional peripheral districts. One of the first actions carried out by the city council after the restoration of democracy was to acquire these spaces.

9.6
The municipal abattoir, site of Joan Miró Park

9.7
Park of the Estació del Nord

Under the new policy, in which public spaces were the most important factor in urban regeneration, the factories were demolished and the sites converted into public spaces, parks where green and architecture are the key elements in urban design and organization.

Due to all this, these parks have special characteristics that distinguish them from parks in other European cities. It should be remembered that the average size is just 40,000 m², very small for a park. The project for such spaces has to redefine and regenerate the location. The architecture surrounding them is poor in quality and in many cases made up of clear examples of the speculation in the periphery during the so-called "development" period. It follows that we can only improve such places by deploying highly intensive projects there. These new parks, though limited, are multi-purpose spaces, and their internal structure is divided in such a way that it is the design that reflects the diversity of uses. We will always find a garden, a place for walking, for sitting in the sun or for children's games ... and where vegetation predominates, a square (that is, a harder area where people can meet) and, finally, a building with public facilities. The park, then, is a space where activities are concentrated.

This style was not pre-established, but its constant employment has led to the definition of an urban model that is both appreciated and the object of study, and which underlies the main projects in the late 1980s and early 1990s in Barcelona. These are also interventions aimed at satisfying the city's needs and remedying its deficiencies after a recent history full of contradictions. Architecture was given complete freedom of movement, producing such parks as Joan Miró (Figure 9.6 and colour plate 8), Espanya Industrial, Clot, Pegaso, Creueta del Coll and Estació del Nord (Figure 9.7).

9.8
Princep de Girona Park

9.9
Diagonal Mar Park, and flats, completed 2002

This process of urban improvement and the introduction of green spaces in the city through the rehabilitation and conversion of medium-sized sites has now been developed into a highly effective method for creating quality spaces, as can be seen in the Olga Sacharoff and Príncep de Girona gardens (Figure 9.8) and the planned new Diagonal-Pere IV Park. Finally, a question hangs over the future, as this way of building and seeing the city has changed in various forthcoming projects. The Diagonal Mar project may prove the most controversial example, and it will be necessary to give it time and much thought once it is completed (Figure 9.9).

Privatization of public space: Diagonal Mar (*Zaida Muxí*)

Diagonal Mar is a property development project in Barcelona that includes a residential area made up of a large park with eight tall housing blocks, arranged in pairs (Figure 9.10 and colour plate 9). Together with a lower block, these form four trapezoidal microblocks that close around a gardened interior with swimming pools. The park is public, in fact, but the sales staff for the flats deny this, telling potential buyers that it only looks like a public space.

However that may be, with the first construction phase of the park and the first pair of tower blocks completed (in September 2002) there is no doubt that this is a public park. It is, nonetheless, obvious that the private developers applied pressure or put forward "proposals" for slicing up this large park, the third-biggest in Barcelona, so that the four residential complexes could be located in it, immersed in the centre so that it is difficult to gain a perception of its true scale. The towers are separated from the park, moreover, by fences and semi-private roads which are open during the day but become the towers' domain at night. That the towers exercise a certain control over life in the park is evident.

9.10
Diagonal Mar Park and flats

9.11
The new Diagonal being opened up in 1998

Besides the towers, the facilities also include a shopping centre (now completed), a hotel and a convention centre. Building continues apace on the latter two to ensure that they are ready for 2004, the next important date on the city's agenda of events. Each element is enclosed within its own functional mechanics, denying urban relations via the street, except as a mere corridor for traffic.

Diagonal Mar is a landmark project in the new late-rationalist urban model that is being applied in the city of Barcelona. This model is made up of partial, fragmentary urbanism formed by isolated urban objects, and is helping to shape the eastern limits of the city (Figure 9.11).

This fragmentary urbanism aims to segregate and fragment the city, privatizing the public space; it is presented as a solution to the dangers of urban heterogeneity and has been successfully applied in several cities, such as Sao Paulo, Mexico City and Buenos Aires, either in the form of blocks or isolated homes, but all inside a strongly-guarded enclosure.

The aim of all this is to create "sleeping" cities inhabited by equals, living in a kind of fantasy Disney World (one thinks of the film *The Truman Show*) in which everything is planned, settled, everyone knows everyone else and all are equals. But a space for equals is not a city. This is a proposal that denies the very essence of the city, which lies in heterogeneity. The city is a place for casual, chance encounter, for getting to know the other with the possibility of conflict and of coexistence. The proposal is also an urban conception alien to the history and spirit of the Mediterranean and European city, which has contributed, fundamentally, a way of collectively using and enjoying the urban space to traditions of urban development. As long ago as late eighteenth-century Italy, which Goethe visited and portrayed in his book *Travels in Italy*, the public right to use all open spaces in the city was defended by citizens, who occupied arcades, galleries, doorways, courtyards, cloisters and the interiors of churches. Mediterranean cities grew up in a wise combination of domestic spaces and public buildings, squares and streets providing access to spaces that form a gradual transition from the public to the private, ambiguous places where the presence of strangers is tolerated. Moreover, the Latin languages gave us the names of many architectural and urban elements devoted to human relations: atrium, peristyle, patio, veranda, porch, vestibule, terrace, belvedere, boulevard, etc., not to mention the café, the city meeting-place *par excellence*.[3]

The emergence of this urban development style in Barcelona is at least surprising, as this is a city proud of its streets and its urban history, as well as a model for the revival of urban life in many cities over recent decades. The idea of private streets and parks is in complete contradiction with the definition of the elements that form the public space *par excellence*: the street and the square.

Notes

1 Cenzatti, M. and Crawford, M. (1993) "Espacios públicos y mundos paralelos", *Casabella*, pp. 597–98.
2 See Various authors (1983) *Estudi de l'Eixample* (Study of the Eixample District), Barcelona City Council.
3 Montaner, J. M. "La calla privida" ("The private street"), *El País Cataluña*, 31 March 1998.

PART 2

Present and futures

Both reality and policy change fast in Barcelona. This part is a condensed look at current developments and future plans for the city. The first chapter is an edited version of a publication of Barcelona Regional of 1999. Barcelona Regional, a multi-person firm, with 100% public capital, was founded in 1993 under the format of a limited company, created by entities and companies from the public sector with the aim of providing a common technical instrument, stable in nature, that would enable cooperation between the different public agents acting in the metropolitan area of Barcelona. It is set up as the shareholders' direct management body and its main function is to provide specialist services to its members and other public agents linked to or dependent on areas of urban planning, the environment and infrastructures of all kinds. The chapter also contains an appendix on financing of the Olympics, to give some comparison with current projects.

Naturally, changes have occurred since 1999. The project for the 2004 Forum is at the time of writing rapidly advancing towards completion, and the 22@ scheme is fully launched, as described and illustrated in the following chapter. However, the chapter does show clearly the overall schema within which the current drives fit.

Chapter 11 examines just one other dimension: the plans for the 22@ zone. It is written by Oriol Clos, an architect and the director appointed by the council to manage the physical and architectural components of the new district.

Chapter 10

Barcelona's new projects

Barcelona Regional S.A.[1]

Barcelona's new projects: a summary

Barcelona's new strategic projects are concentrated in the area of the Besòs and Llobregat rivers, which flow respectively through the northern and southern ends of Barcelona. The projects for the Besòs area focus on urban and environmental renovation, while those in the Llobregat area concern logistics, transport and productive functions, and environmental issues as well (colour plate 10). All the projects are conceived within the much wider Bracelona region (Figure 10.1), which itself has to be understood within the situation of Catalonia within Europe (Figure 10.2).

The Llobregat river area
Productive projects in this area include:

- enlargement of the airport (construction of a third runway and terminals);
- development of the "Airport city" (cargo, maintenance and service areas);
- enlargement of the Port of Barcelona, which involves diverting the final stretch of the river Llobregat;
- development of the logistics activities area (ZAL) related to the port;
- Zona Franca Consortium Logistic Park ("Free Port Consortium");
- enlargement of the trade fair grounds (Barcelona Fair II).

10.1
Urbanized central areas of Barcelona metropolitan region

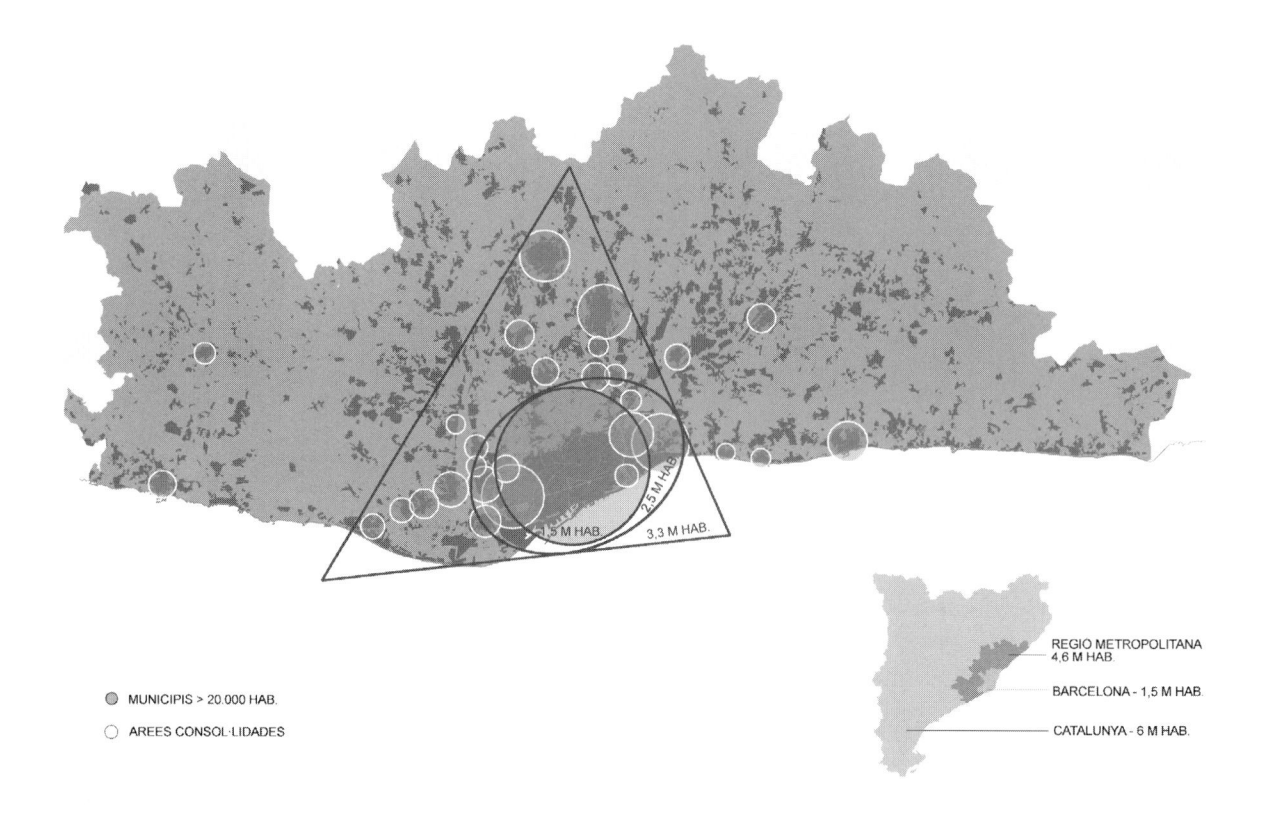

Environmental improvements for this area include:

- wetland conservation in the Llobregat river delta;
- coastal regeneration;
- sewage treatment plant;
- Ecopark solid waste treatment area.

The Besòs river area

Along the Barcelona–Girona–France rail corridor:

- new inter-modal station at Sagrera (TGV, regional, suburban, underground rail lines and bus facilities) and new residential, commercial and business areas linked to the rail corridor.

Along the Avenida Diagonal (main boulevard running the length of the city):

- extending the Av. Diagonal to the sea; construction of new residential areas and location of headquarters offices;
- new sea front — residential;
- new shopping and business centre.

10.2
Levels of GNP per capita in Europe in 1996

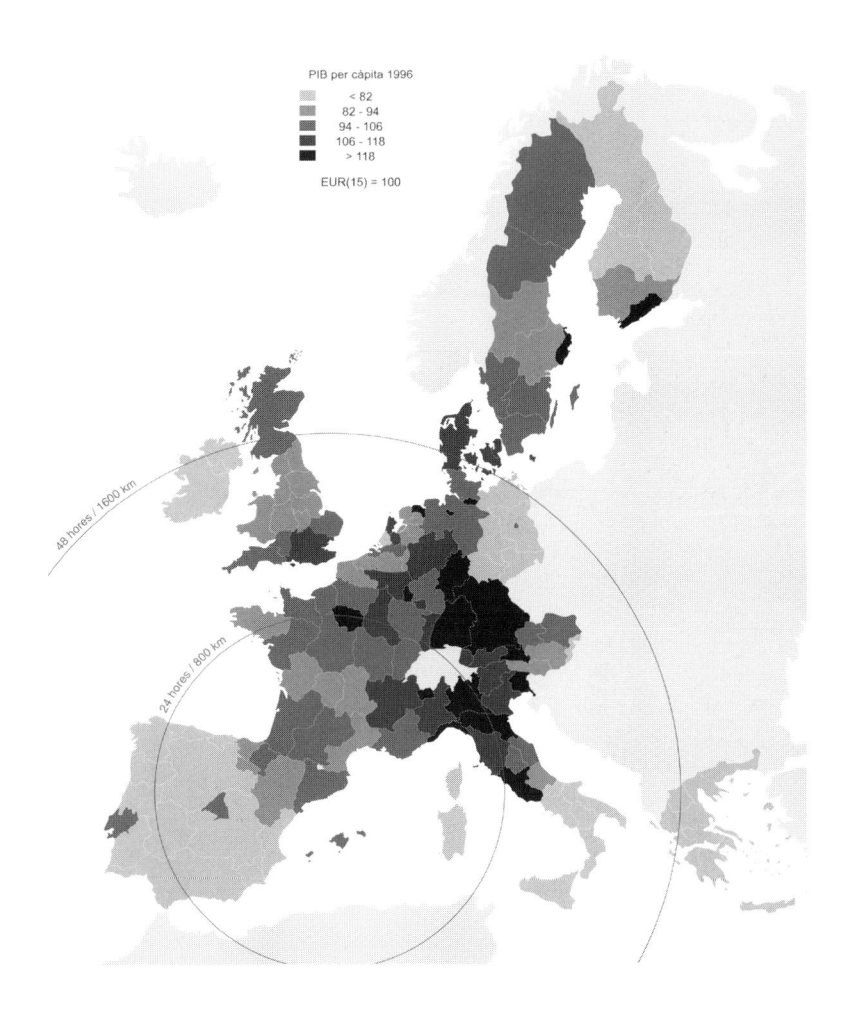

In the Poblenou area:

- renewal of an old industrial area for the location of "advanced industries" and offices (22@ — see Chapter 11).

Along the Besòs river:

- rehabilitation of the urban stretch of the river, including work to restore the natural river bed.

Along the coastline (waterfront):

- completion of the major transformation of the coastal zone of Barcelona that was begun in 1992. Regeneration of the coastline, new public spaces and facilities, construction of a new central urban area equipped with an international convention centre, hotels and office spaces.

BARCELONA'S NEW PROJECTS **177**

The importance of infrastructures in establishing complex business districts: the port, the airport and the logistics areas in the Llobregat delta

The Llobregat Delta Plan comprises a set of investments and coordinated measures for transport infrastructure and environmental protection, forming part of MOPTMA's (Ministry of Public Works) Infrastructure Master Plan 1993–2007, which was published in 1993.

The Delta Plan, which envisages an initial investment of 400,000 million pesetas (2,410 million euros) is currently under development (colour plate 11). Agreements on the project were reached between the Spanish and Catalan governments and local administrations in 1994. The Delta Plan includes the following schemes:

- extension of the port and diversion of the river Llobregat;
- logistics activities area (ZAL);
- extension of Barcelona Airport;
- road and rail infrastructures, forming part of the metropolitan and regional networks;
- waste water treatment plant and solid waste treatment area — the latter will allow closure of the Garraf municipal rubbish tip;
- environmental management in the wetlands and in regeneration of the coast.

The Delta Plan infrastructures will promote the activity of nearby industrial estates, as well as those areas dedicated to logistics activities. First, the Free Port industrial estate, with an area of roughly 600 hectares and providing 30,000 jobs is one of the biggest in Spain (the SEAT car factory, covering 150 hectares is currently being converted for logistical uses). Second, the Pedrosa, Gran Vía Sud, and Passeig de la Zona Franca industrial estates sited in the Barcelona and Hospitalet municipalities enjoy a central location and will attract industry and tertiary activities.

Finally, Mas Blau I (34 hectares), with 60% of building work completed and Mas Blau II (74 hectares) of which 130,000 m² of floor space has been earmarked for airport activities, will provide a large reserve of land for those activities best placed to take advantage of planned infrastructure improvements.

Port: entry point to southern Europe and the logistic area (ZAL)/ Free Port

The Port of Barcelona is the biggest port in the Mediterranean in terms of number of vehicles transported. It handled over 24.9 million tons of freight in 1997. These figures are still modest compared with the 478 million tons handled by North Sea ports such as Hamburg, Bremerhaven, Rotterdam and Antwerp which serve the Atlantic trade. The Port of Barcelona currently covers 542 hectares. The extension of the logistic area, made possible by diverting the river and reclaiming land from the sea, will make 773 hectares available in the year 2000. The Master Plan envisages that the port will reach its maximum size in the year 2010 when it will cover 1051 hectares, 1317 including logistics area achieved thanks to extension of the port's enclosing dykes.

The past few years have witnessed the consolidation of the old port area as an integral part of the city, now free of the goods which once cluttered the city's waterfront. This transformation will be completed with the renovation of the ferry passenger terminal, the completion of an office and service building (World Trade Centre), and a second harbour mouth which will facilitate the separation of passenger vessels and private leisure craft.

The consolidation of the logistics area (ZAL) is required to ensure the Port of Barcelona remains the main distribution centre for southern Europe. The coordination of the remaining elements making up the Delta Plan and their direct link to metropolitan and other infrastructures (particularly the Free Port and freight depot at Santa Perpètua and the Azuqueca freight facilities in Madrid) are yet further measures aimed at achieving this goal. The logistics area (ZAL) currently occupies 65 hectares. This will reach 268 hectares once the river Llobregat has been diverted along its final stretch. Once the extension work has been completed, it is forecast that 15,000 people will work in the facility which should move 9 million tons of freight a year.

Airport: hinterland requirements for a traffic volume of 40 million passengers a year

Barcelona Airport ranks seventeenth in Europe in terms of passenger volume. Annual growth has averaged 8% since 1985. Now passenger numbers exceed 20 million a year and the freight volume is about 60,000 tons per year. The rate of passenger growth has been 7.4% (1997–98), 7.5% (1999–98), 13.7% (1999–2000).

Increasing specialization is envisaged, both in terms of passenger and freight traffic, with the latter playing an increasingly important role. Fully 25% of world trade in manufactured articles by value is now transported by air, whereas in Catalonia this only represents 5% of all imports and exports.

The scenario for airport development is one of growth in passenger traffic, reaching over 30 million passengers per year in 2011 and 40 million in 2016, assuming steady annual growth of 6%. Barcelona Airport will not be able to function as an international hub when it reaches such a traffic volume. With regard to freight, it is reasonable to expect that it will handle half a million tons of freight by 2020. Achieving these passenger and freight volumes will require adequate services, capacity and airport size.

The increase in traffic will involve growth in industrial activities, particularly those related to aircraft maintenance and related fields. Barcelona Airport currently provides 7,500 jobs or 500 jobs per million passengers. With an increase in airport functions, this could reach a level of 1000 jobs per million passengers. Applying this yardstick would mean the airport could provide some 30,000 jobs, located in the new industrial floorspace (700,000 m² approximately) as well as in the terminals. The airport's impact on employment in the El Prat municipality is already considerable: it currently accounts for 20% of jobs in the area, while 42% of airport workers reside in the borough.

In order to reach these objectives it will be necessary to build the installations envisaged in the Master Plan:

- runways — a third runway is required. This will have to reconcile airfield operating requirements with environmental considerations (there are important wetlands and nature reserves in the area);
- reception points for planes;
- terminals — up to 585,000 m² of buildings (110,000 m² currently in service);
- internal connections with an underground "people mover";
- freight loading area on the land side, covering 59 hectares, providing facilities for logistic activities;
- service area on the air side, with an aircraft maintenance area covering 7 hectares;
- general maintenance and service area covering 80 hectares;
- "Airport city" providing ancillary services and activities. A central area of 100 hectares is planned for this purpose.

The airport's rail station will have six lines and be re-sited below the future terminal building. The station is a keystone in the strategy for achieving traffic growth. It would be very difficult to serve demand for air transport in the city's 300–400 km hinterland without complete integration with the railway network in general and a high-speed rail link to the French border and Madrid in particular.

Barcelona — new shoreline next to the river Besòs: environmental infrastructure and the 2004 Forum

The plan for the Besòs area comprises a set of projects covering a site of over 200 hectares (colour plate 12). The Besòs area is in the far north-eastern corner of the city and is bounded by the shoreline and the right bank of the river Besòs itself.

The scheme will round off the first great change to the Barcelona coastal strip on which work began in 1992. This time, remodelling work will take place in the Barcelona and Sant Adrià del Besòs municipalities.

The public will gain access to a wide range of new spaces. The project's general aims also include renovation and rezoning of the neighbourhoods in the area (colour plate 13).

The following measures have been proposed to accomplish these aims.

Creation of a specific management body

A management body is needed to ensure efficient execution of works and proper coordination of schemes and the private and public bodies involved in the project.

This managing body already exists in the form of the Consorci del Besòs (Besòs Planning Consortium), formed by the city councils of Barcelona and Sant Adrià del Besòs for this purpose. The consortium coordinates and assists in drawing up and approving outline plans.

Barcelona Regional, Agència Metropolitana de Desenvolupament Urbanístic i d'Infrastructures S.A. (Barcelona Metropolitan Planning Agency Ltd) has been charged with the technical management and direct definition of the Master Plan and preliminary projects.

The elaboration of the executive projects has been jointly coordinated by Barcelona Regional and Infrastructures de Llevant de Barcelona S.A., a company set up specifically to carry these projects through to their conclusion.

New sea front: Universal Forum of Cultures — Barcelona 2004

The Besòs area of the city has been chosen to celebrate the 2004 Universal Forum of Cultures. The event will give a fillip to the town planning schemes needed to transform an area with considerable potential.

The Universal Forum of Cultures — Barcelona 2004 is a world event, conceived as an international gathering to discuss issues of peace, cultural diversity and environmental sustainability as mankind enters the twenty-first century.

The 2004 Forum stresses sustainable urban development. The facilities built for the event (Convention Centre, fairground, water park, etc.) will subsequently be put to good use by the city. Accordingly, the construction work for the Forum will lay emphasis on environmental rehabilitation schemes for restoring the sea front and the Besòs river bed and banks.

A new shoreline is planned (colour plate 14). Part of the shoreline will be built on reclaimed land in order to make space for the new city Zoo-Park. The yacht haven will provide 1,000 moorings, with associated business and hotel activities. There will also be a bathing area and a sea front park. The

10.3
View of proposed coastal changes, looking south west along Diagonal

promenade will bridge the river Besòs thus linking the Sant Adrià and Badalona sea fronts. The prolongation of the Diagonal (main city thoroughfare) to the inner harbour of the new Port of Sant Adrià is intended to make the remodelled shoreline an integral part of the city (Figure 10.3).

The foregoing proposals for the new seashore will also incorporate environmental considerations with regard to wetlands, marshes, reefs and improving the biodiversity of the sea bed.

Technical infrastructure in the metropolitan and river Besòs areas
Planned environmental rehabilitation work along the seashore and the final stretch of the river Besòs (currently severely blighted) includes various technical projects and changes to existing installations (waste water treatment plant, solid urban waste treatment and electricity generation). These measures are required if environmental standards are to be met. The gradual technical improvement of these installations also forms part of the overall programme. Schemes forming part of the programme and now underway include: restoration of the river bed; the building of a sewage sludge drying plant; and the installation of new incinerator filters.

The burial of high-voltage electricity lines is planned for the near future, as is the incorporation of co-generation units and introduction of secondary treatment of waste water.

Llull-Taulat: a new residential and business district neighbourhood
A new residential district is planned. The new neighbourhood will incorporate sustainable development and energy-saving criteria. The scheme will

comprise some 1200–1500 dwellings and include industrial estates, and areas for siting tertiary activities and hotels, complementary to the residential nature of the project.

These dwellings will be designed and priced to attract young middle-class people and thus provide a shot in the arm for an area where the population has aged over the past 25 years.

The area will also incorporate new housing technologies and provide a proving ground for urban sustainability projects linked to the Universal Forum of Cultures — 2004.

Bordering residential areas such as "La Mina"

Rehabilitation work is also planned on the blighted bordering areas. A combination of environmental rehabilitation and schemes in the neighbourhoods themselves in the short term and rehabilitation of dwellings in the medium term will bring these run-down areas up to decent town planning standards. Eventually, a partial remodelling could be implemented.

"La Catalana"

The "La Catalana" area is located in the Sant Adrià del Besòs municipality and is included in the various measures planned for the area. It is strategically positioned on the right bank of the river Besòs and is surrounded by local industries which are currently undergoing modernization and rationalization.

"La Catalana" is a residential neighbourhood, which has gradually lost population over the years. There are currently 80 families living on the site. They will have to move out if urban renewal plans for the area are to be carried out. Part of the land is publicly owned, thus making execution of the scheme much simpler.

"La Catalana" has excellent communication potential:

- *regional and metropolitan links*: direct links to the coastal ring road and the A-19 Motorway (the main link between Barcelona and Maresme area — one of the most dynamic metropolitan areas), as well as a suburban railway line connected to the future intermodal TGV station at Sagrera;
- *urban links*: the Avenida Cristóbal de Moura (major road) links the historic city centre with urban areas along the left bank of the river Besòs.

The planned urban renovation work for the "La Catalana" area could envisage several options. However, all of them have a common purpose: the creation of an area with a pleasant landscape and attractive environmental features (an important point since the site is close to the mouth of the river Besòs). Proposals include:

- a university or technical training campus for the metropolitan area;
- a technology park (research and business activities);
- a residential area.

The final decision will be based on thorough financial and technical studies. The site covers some 25 hectares. Current plans envisage a built area covering some 110,000 m², most of which would be sited in parkland.

Sagrera–Sant Andreu: remodelling of the urban environment and the future central railway station

The construction of the high-speed train (AVE) from Madrid to the French border and its arrival in Barcelona in 2004 justifies a new central station in La Sagrera. A new AVE station connecting this line to the airport is also critical to reach its target of 30 million passengers. The flexibility of the network is guaranteed if Sagrera is not a terminal point and this will be possible with the connection Airport–Sagera.

The Sagrera station could be of considerable importance for the city and also for the northern municipalities. The Sagrera–Sant Andreu project involves remodelling rail traffic running through the northern gateway to Barcelona. The scheme will be rounded off by large-scale planning changes, including a linear park over 3.5 km long. The whole area will cover nearly 1.5 million m² of floorspace for housing, facilities and offices.

The new station will provide the railway services of suburban, regional and high-speed links (both medium and long haul). Each of these rail services makes very different demands on infrastructures. Accordingly, separation of the different types of rail traffic will be required.

Sagrera will be one of the key points in ensuring proper interchange between transport nodes and networks for passengers, in addition to airport passenger traffic (high speed, regional and suburban trains).

It will complement freight traffic coming from port and logistic areas in the Llobregat delta (ZAL, Free Port, Can Tunis, Morrot), which will run in the Llobregat corridor (Spanish, international and metric gauges).

With completion of the high-speed rail network there would be over 150 train movements (i.e. departures and arrivals) a day covering all types of traffic. Some 750 trains per day (150 trains on the high-speed rail network and 600 on the regional rail network) and 30 million passengers per year would use the new rail station, which emphasizes the scale of the impacts of Sagrera station on the city.

The priority accorded Sagrera station is fully consistent with the stated aim of making sure trains reach the city centre. This is precisely one of the big advantages rail enjoys over other forms of transport, opening at the same time important opportunities for remodelling the urban environment in its surroundings.

Finally there is the 22@ project. This is covered in detail in the following chapter.

Financing metropolitan infrastructure in Barcelona: cash flows and public budgets

The infrastructure projects (excluding associated real estate operations) represent a total investment of 1.9 thousand million pesetas in the Barcelona Metropolitan Region. Tables 10.1–10.3 on the following pages provide a breakdown by sector and type of funding. Table 10.1 is a projected overview from 1999, whilst Tables 10.2 and 10.3 are from 2002, giving data in relation to the 2004 phase of work as it is actually being implemented. The comments below are based on the forward view of 1999.

The following aspects should particularly be noted.

- A quarter of the investment required is in the public transport sector, especially for the new underground network (720,000 million pesetas).
- The estimate for the high-speed rail link includes the pro rata figure for the Sagrera Central Station. The investment in the TGV amounts to 218,000 million pesetas in the Metropolitan Region alone.
- The extension of the port and airport, together with the high-speed train project provides the key to realizing the potential of Barcelona's logistic industry and improving communications with the rest of Europe. These investments (330,000 million pesetas) represent 15% of the total.
- Investment in telecommunications includes cabling the Metropolitan Region with optical fibre. This, together with investments by telephone operators, amount to an investment of 323,000 million pesetas (17% of the total). Telecommunications and transport infrastructure (the ring road especially) are the two fields attracting the biggest investments, amounting to 41% of the public total.
- Priorities set for 2004 represent 60% of total forecast investment.
- 47% of priority investments for 2004 can be financed by cash flows generated by the infrastructure itself (telecommunications, power grid, airport, port and one-third of the high-speed train). The figure would be 43% of the total investment when a grant for the port is taken into account.
- Total investment will amount to 14.2% of 1998 regional GDP over a ten-year period. (1.4% per year). This is a considerable

Table 10.1 Barcelona Metropolitan Region investment — priorities for 2004 and investment profiles (millions of pesetas), and metropolitan infrastructure (millions of pesetas, present value); tables prepared in 1999

Projects	Total investment Millions of pesetas	%	% regional GDP	Programme 2004 (millions of pesetas)
High-speed rail link	218,500	10	1.4	90,000
Conventional rail link	111,500	5	0.7	54,000
Motorways	224,000	10	1.5	74,000
Telecommunications	323,000	15	2.1	323,000
Electricity grid	98,000	4	0.6	98,000
Port extension	123,000	6	0.8	109,000
Airport extension	207,000	9	1.3	117,000
Environmental infrastructure	158,000	7	1.0	135,000
Metropolitan public transport	720,000	33	4.7	432,000
Total	2,183,000	100	14.2	1,432,000

Programme 2004/total investment: 66%

	Programme 2004 (millions of pesetas) From revenue	By public authorities	% funding From revenue	By public authorities
High-speed rail link	30,000	60,000	33	67
Conventional rail link		54,000		100
Motorways		74,000		100
Telecommunications	323,000		100	
Electricity grid	98,000		100	
Port extension	49,000	60,000	45	55
Airport extension	117,000		100	
Environmental infrastructure		135,000		100
Metropolitan public transport		432,000		100
Total	617,000	815,000	43	57
Proportions of non-subsidized port funding:			47%	53%

Comparison with 1992 Olympic Games (in millions of pesetas — 1992):
Total cost 1987–92   1,150,000
% regional GDP 1987–92 10.3%
% regional GDP by year 1.7%

commitment by both the business community and local and national government. It seems a reasonable burden to assume when one considers organizing the Olympic Games represented combined public and private investment equivalent to 10.3% of regional GDP over a 5–6 year period (1.7–2% per year).

Table 10.2 **Infrastructure investment projected in relation to 2004 Project**

	€ million	%	Infrastructure
Central Government and EU funds	150.85	8.7	Public transport improvements and sewage treatment plant modernization
Autonomous Government	241.61	13.9	
Local Government and EU funds	634.67	36.5	Public spaces, roads and squares, coastal reclamation
University and R&D	72.12	4.1	University buildings
Private sector	640.08	36.8	Hotel, offices, housing and burial of high-voltage lines
Total	1,739.3	100	

Source: *Infraestructures 2004*

Table 10.3 **PDI 2001–2010 investment projected under infrastructure plan for Barcelona Metropolitan Region 2001–2010**

Network	Cost: € million	Length: km	Stations
Underground	3,940	62	63
FGC (regional government rail)	1,419	34	29
RENFE (national government rail)	908	0	11
Light rail	429	31	65
Others	359	21	51
Total	7,055	148	219

Source: ATM (April 2002), Barcelona Regional (June 2002)

However, planned public investment in metropolitan infrastructure for 2004 comes to 1% of Catalonia's GDP, which will mainly be borne by the Catalan and Spanish governments. This has to be set against the backdrop of a progressive fall in public investment, which took place from 1991–97. The Central Government invested less in Catalonia (0.92% GDP) than in the whole of Spain (1.8% GDP), the difference (i.e. 0.8 % of GDP) being equivalent to 111,000 million pesetas at present values (669 million euros).

The precise effect of including regional and local government investment in the difference mentioned above varies according to the data used. According to the Chamber of Civil Engineering Contractors, it amounts to 0.37% of GDP, so that local government's investment would be partially offset by central government investment. However, according to Prof. Núria Bosch, the figure would rise to 1.2% of GDP. According to this second estimate, there would be a negative investment difference per inhabitant between Catalonia and Spain, amounting to 49,000 million pesetas (some 8000 pesetas per capita per year).

In this respect, it should be pointed out that failure to redress infrastructure shortcomings through rational investment will put a brake on future economic growth. A recent study by the GT-Y group (comprising the Chamber of Trade, Ministry for Employment Creation, Economic Circle, and the Royal Catalan Automobile Club) pointed out that "Madrid, Catalonia, the Basque Country and the Valencia Community suffer the greatest infrastructure deficits but where returns on investments in infrastructure are highest".

Appendix
The 1986–92 investment programme for the Olympic Games as a model: main figures

Funding and management implications of the Olympic Games
There are a number of useful planning lessons to be learned from the management and funding of the Olympic Games. The following section provides more detailed figures. The salient points can be summarized as follows.

- Addressing the main *infrastructure deficits* in the "Metropolitan City" of the 1980s concerned the ring roads (B-20 and B-30), the airport, telecommunications and hotels.
- A *world event* like the Olympic Games was useful as a means of concentrating public and private investment. The event itself only represented 15% of total expenditure and "agreed" investments (1.1 thousand million pesetas). Sports facilities only accounted for 9% of total expenditure.
- *Barcelona City Council* played a leading role in planning projects and ensuring high standards were met in their implementation. Fully 39% of investment was in the Barcelona municipality, 34% in the rest of the Metropolitan Region, whilst 25% covered general infrastructure (telecommunications, etc.).
- *Collaboration between the private and public sectors* was important, since public funding would have proved insufficient by itself. In monetary terms, the public sector managed 65% of investment, although its net investment was only 45% of the total since the private sector funded dwellings in the Olympic Villages (all of the 4,000 dwellings were finally sold in 1995, of which 2,500 were in the Olympic Village). The same applied to telecommunications facilities, hotels and offices.
- There was institutional consensus on making 450,000 million pesetas' worth of public investments, split between the Spanish government (41%), the Catalan government (32%), Barcelona City Council (18%) and provincial and metropolitan administrations (9%).
- A management formula was developed between the Spanish and Catalan governments which established *Holding Olímpico* (Olympic Holding Company). The holding company managed 43% of public sector investment, including the Olympic Village.
- *Investments had a regional impact* (1.8% of Catalonia's GDP over six years), creating 128,000 man-year jobs or approximately 30% of total employment created in the city.
- Total investment also included "*underground infrastructure*" such as sewers, storm drains and service galleries along the 35 km of new ring roads.
- The Olympic project was integrated with the broader *Barcelona 2000 Strategic Plan (1994–1998)*, which looked beyond the event itself ("the day after"). Transport infrastructure represented 42% of total investment.

- Barcelona's historic city centre received a considerable slice of this: 38,000 million pesetas of municipal investment over six years, while Olympic investments received 79,000 million pesetas.
- Some of the benefits from these investments appeared years later. There were 4.1 million overnight stays by tourists in 1991. This figure rose to 6.3 million in 1995, with American visitors making up the most important group. The number of 3, 4 or 5 star hotel rooms was 10,812 in 1991, which grew to 15,076 rooms in 1995.
- *Investment in the Olympic Games was set at a "reasonable" level.* The City Council's and Catalan government's investments together represented 18–20% of their total investment in the period between 1986 and 1992, while the Spanish government improved on its past investment in Catalonia by 40% over a six-year period.
- *Every peseta spent by the City Council was effectively matched by 14 pesetas from other sources.*

Both the effort put in and the circumstances were unique. However, the Forum of Cultures — Barcelona 2004 aims to work on the same lines. *Cities need integrated, strategic projects* if they are to develop. Future metropolitan infrastructure projects in Barcelona have been estimated at 2.2 thousand million pesetas.

Main figures

- Total cost US$9,376 million (1,100,000 million pesetas).
- *Comparative figures* for other Olympic Games in millions of US dollars; Seoul 88 (3,155); Los Angeles 84 (522); Montreal 76 (3,175); Tokyo 64 (16,826). Only Tokyo spent more than Barcelona.
- Organization and holding of the *Olympic Games: 15% of the total cost (162,000 million pesetas)*: 1.8% of GDP over six years (10.7% of the 1991 GDP).
- *Olympic Committee expenses: 196,000 million pesetas*, including 17% spent on sports facilities.
- Investments in *infrastructure and buildings*: 85% of total expenditure (957.000 million pesetas).
- *Economic importance of total expenditure* in relation to GDP (Catalonia's 1991 GDP: 10.3 billion pesetas): 1.8% over six years (10.7% of 1991 GDP).
- *Sectoral breakdown* of investment: roads (42%), including service galleries; housing and offices (15%); hotels (13%); telecommunications (13%); and others (17%).
- *Publicly managed funds*: 72% of total expenditure (807,000 million pesetas) and *67% of investment in infrastructure* (644,000 million pesetas).
- *Public sector funding* (budget general resources = sources of funds): 40% of total expenditure and *47% of infrastructure investments* (450,000 million pesetas). The remaining 53% of investments were funded by proceeds from real estate asset sales, road tolls, etc.
- *Olympic Holding* (HOLSA) managed (partnership between the City Council and the Spanish government): *275,000 million pesetas — 43%* of total publicly-managed expenditure.
- *Funding from the Barcelona Municipal Budget: 79,100 million pesetas —* 18% of total public funding (23,000 million pesetas in the 1987–92 budgets).
- *The Olympic debt burden borne by the Barcelona Municipal Budget: approximately 20% of total investment capacity over 20 years —* up to

1992: 20.3% (23,000/19,000 × 6); 1993–2007: 21.4% (7,900 × 15)/(36.877 × 15)(1996:22.9%).
- *Multiplier effects of municipal investments* and plans: 5.7:1 in public investment, 14:1 in total direct expenditure.
- *Spanish government funding: 185,000 million pesetas* — 41% of total public funding: this represented a 35% increase over central government investment in Catalonia between 1987 and 1992.
- *Regional government funding: 143,000 million pesetas* — 32% of the total public investment burden: 18% of the regional government's total investment between 1986 and 1991.
- *Total impact on the Catalan economy*: 3:1 multiplier, creating *21,000 jobs over six years*.[2]

Note

[1] This chapter was updated by Maria Buhigas of Barcelona Regional.
[2] Data sources: Brunet, F. (1994) "*Economia do los Juegos Olímpicos de Barcelona*" I.O.C. Lausanne, pp. 107–90. Versions in English, French and Spanish and extracted information.

Chapter 11

The transformation of Poblenou: the new 22@ District

Oriol Clos

Origins and past of the 22@ Area

Poblenou was the coastal sector of the former municipality of Sant Martí de Provençals, incorporated into the City of Barcelona in 1897. The area is built on land reclaimed from the old coastal lagoons to the south of the river Besós delta. In the second half of the nineteenth century an urban centre, basically of an industrial nature, was established there, constituting a new district of Barcelona to the east of the old city, on the far side of the military citadel and railway lines. Up until the time of the large-scale infrastructure work for the 1992 Olympic Games it had remained cut off from the central part of the city by substantial physical barriers.

The earliest modern industries of Barcelona were established in the district from 1850 onwards, taking full advantage of the abundant underground water sources, and the ease of communications with the port, the gateway both for the incoming raw materials and coal, and for the outgoing manufactured goods shipped to foreign markets.

The process of industrialization of Poblenou, so well narrated by Professors Nadal and Tafunell in their study *Sant Martí de Provençals, Pulmó industrial de Barcelona. 1847-1992* (ed. Columna, Barcelona 1992), began with the establishment of the textile sector in the area, first the whitening, printing and finishing sectors which, in turn, with the coming of steam power, later attracted the spinning and weaving mills. Later the food industry, mainly flour mills and alcohols, likewise developed alongside the textile sector and the incipient metal industry, electrically powered, which came to be the dominant industry of the area around the middle of the twentieth century. More recently, at the end of the 1960s, with the relocation and dismantling of much of the existing industry, logistics and transport became the main sectors. Companies set up in the many small workshops that had resulted from the breaking up of the old large industrial sites, in an ever more degraded urban environment, subject to the general decline of the industrial wealth of a district under great pressure to transform, in ways not provided for in the General Metropolitan Plan (PGM), in force since 1976.

The 1859 Eixample (Extension) Plan by Cerdà established the urban layout of the whole of the plain of Barcelona, including Poblenou. From the very beginning the new industrial fabric was laid out in line with the Cerdà plan, alongside the old roadways and existing urban centres, articulated around the road to France (carrer Pere IV). Historically, Pere IV was the most important artery of the area and it was retained as a diagonal running across the new grid system to guarantee access to the whole of the industrial sector and the increasingly important and consolidated residential sector of this former coastal district of Sant Martí. Up to the present, the large size of some of the industrial sites, covering more than one city block, the discontinuities between the district and the remainder of the town caused by the railway lines, and the conflicts between the original nineteenth-century layout and the Cerdà grid, have prevented the opening up of all of the Eixample streets provided for in the 1859 plan.

The successive plans devised for Barcelona from the beginning of the twentieth century, especially those for the latest stages of the city's development (the 1953 County Plan and the 1976 General Metropolitan Plan), established the development criteria required to complete the grid of regular city blocks, and the mono-functional specialization of the industrial fabric of Poblenou. It was not until relatively recent residential operations, starting at the end of the 1980s with the removal of the overground railway lines and the construction of the Olympic Village, that the transformation of the whole of the sea front and the completion of the Diagonal through to the sea permitted Cerdà's century-old plan to be finally completed, with the consolidation of the Eixample grid and its main diagonal roadway, the urban backbone of the sector.

Forms and uses for the 22@ Area transformation

The morphology that has resulted from this long and complex process is characterized by the coexistence of industrial, residential and service sector buildings, of greatly varying size, importance and styles, in very close proximity, and with what is, at times, brutal discontinuities and breaks. A highly irregular fabric is formed, with little homogeneity. There are huge industrial sites, residential tower blocks, virtually separate individual houses, simple single-storey industrial constructions, potent concrete industrial buildings of over five floors, in short, an extensive catalogue of building types, which house such diverse activities as storage, industry of all types, housing, services and local facilities, expressing the wealth of the industrial and residential architectural prose of Barcelona (colour plate 15).

The brick chimneys of the old steam engines and the remaining pieces of high-quality industrial architecture of many different periods are juxtaposed, with no prospect of continuity, along the length of the streets, giving rise to a varied, changing, kaleidoscopic, diverse landscape with a wealth of images, forms and contrasts and great potential to accentuate its own character, so different to that of other parts of the city.

The passages and interstitial spaces to be found throughout the fabric of the blocks, and the arrangement of the buildings around the inner courtyards and open spaces, service areas and ventilation wells, establish a morphology of the city block which diverges from the model of the square of buildings of even height, aligned along the streets, that characterizes the city block of the central Eixample area. These specific formal characteristics deriving from the indiscriminate use of land, based on the logics of productive and residential use, characterize both the industrial blocks of Poblenou, and the form of occupation of the residential blocks of the old part of the district, in all a total of 150 Eixample city blocks of a very distinctive kind.

Amendment of the PGM, aimed at the refurbishment of the industrial areas of Poblenou and the creation of the 22@ District of Activities, was first studied at the end of the 1990s. This amendment was to establish the criteria for the transformation of the land previously classified as industrial land by the PGM (22a classification). At the time, the plots occupied by the old, degraded, obsolete industrial uses were subject to great speculative pressure, seeking to convert them into land classified for residential use. The 22@ Plan establishes the criteria and terms for the conversion of the old industrial areas into a sector suited to the new forms of

11.1
Aerial view of 22@ area in relation to adjoining areas

productive activity, based on information and knowledge technology, and on a new balance of urban, residential, productive and service functions, in which all are integrated into a hybrid fabric, constructed around the historical morphology of the sector. The transformation of the industrial sector of Poblenou is a part of a far larger strategic plan developed for the whole of the eastern side of Barcelona, the area contained by a great triangle, of over 3 km per side, defined by the La Sagrera rail corridor, Rambla Prim and the newly-opened extension to Avinguda Diagonal (Figure 11.1). The area includes the new high-speed inter-modal railway station, the new hub at La Plaça de les Glòries and the newly created urban spaces, alongside the river Besòs, and the infrastructure along the sea front, which is to serve as the venue for the events of the Forum 2004. Thus, the new part of Avinguda Diagonal serves as the backbone for the transformation of the old productive sectors of town.

Processes of controlling change

The 22@ Plan is a regulatory document which covers an area of 115 city blocks of the Barcelona Eixample. It seeks to transform the existing structure to permit the development of new floorspace for production facilities, based on the progressive increase in the accepted levels of use and density permitted, on what had previously been classified as industrial land. It establishes a range of initiatives, of very differing scope, which run from direct action on existing plots, through to large-scale public plans covering areas of more than one city block, which require complex management and land assembly processes, as well as plans for just one block promoted by the private sector. The town planning process also permits an increase in the land used for public services (facilities, open space and social housing), and the installation and renovation of infrastructure, improving both the quality and capacity of urban services, energy supply and telecommunications services, water, waste management, mobility (using all transport modes) and the specialized use of public space.

On the basis of the general regulations, special plans establish the specific pace and form of transformation, including morphological, typological and spatial criteria, adapted to meet the needs of the specific operator and site. These plans, of very differing scope, may be sponsored by either public or private agents, to respond to the broader criteria established for the urban structure of the sector — defining the strategic axes, the larger sites and public facilities, the location of new housing, the volumes of the buildings, and the type and nature of the different kinds of public space. These special plans acknowledge existing social, economic and spatial values and, to a large extent, conserve and enhance the existing fabric, establishing conditions of coexistence that permit the definition of the new urban forms and images that characterize this new central productive district. Thus, it is an operation of internal urban refurbishment, based on isolated, heterogeneous, fragmentary

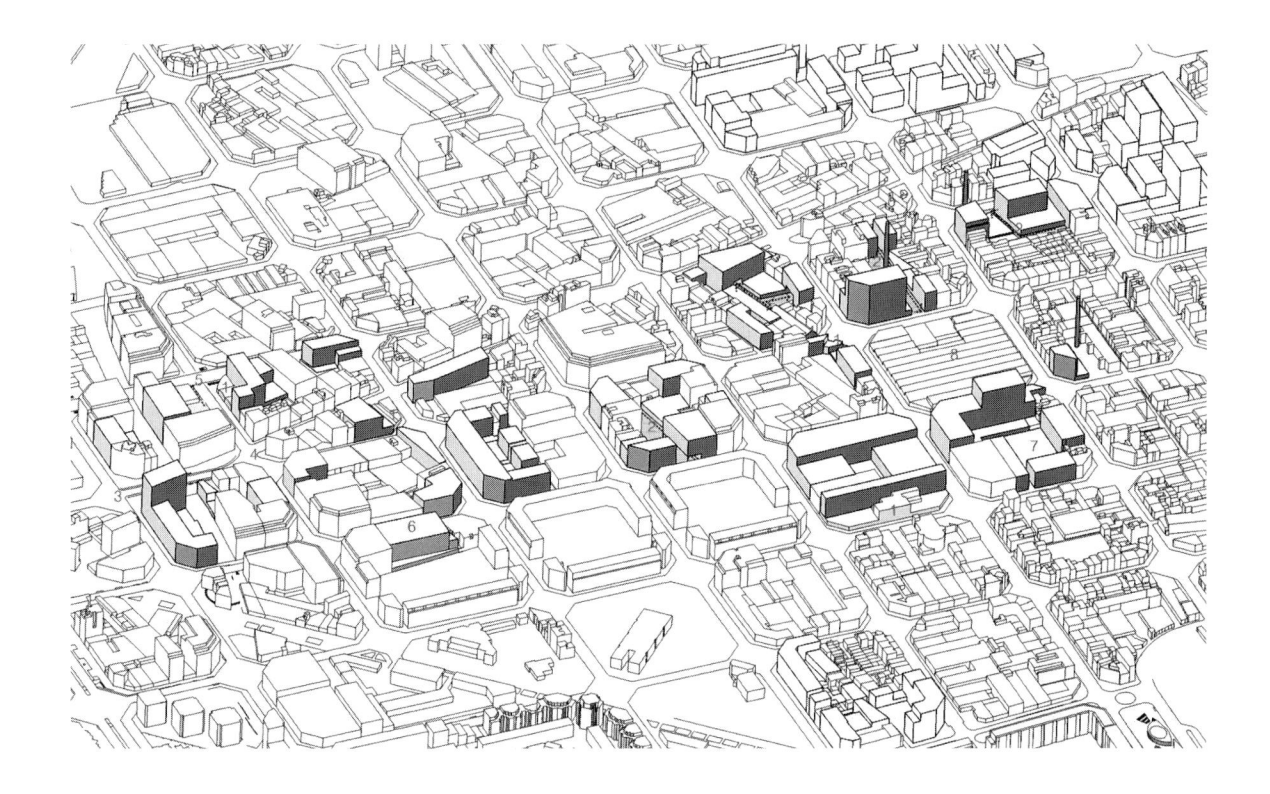

11.2
Perspective view of new blocks proposed in parts of 22@ area

areas of action, and featuring the coexistence of the new and the old which, to a large extent, is conserved. New buildings emerge alongside the reused old buildings, following the existing layout of land use and building which establishes the unique nature of the district. This therefore continues from the morphology of a productive structure which has existed in the area for over 150 years. A new typology of buildings to house offices, urban industry, hotels, services and facilities, all directly related to the new uses, is established on the basis of extremely simple relationships of contiguity and juxtaposition, coexisting alongside the existing fabric which is either to be preserved or is awaiting transformation (Figure 11.2). The new buildings ought to provide a large part of the new productive floorspace and give the district a new form of urban spatiality. This will be based on interstitial public space blended with the powerful structure of the Eixample grid, configuring new scenarios which superimpose close and distant perspectives and form new urban axes, distinctive sites and a fabric which is functionally mixed and formally complex.

Appendices from the leaflet for diffusion (January 2001)

Uses permitted in New Zoning Regulations

The resulting instrument for bringing about the transformation of this area is the Modification of the (PGM) passed on 27 July 2000. The Modification departs from the emphatic concept of land specialization as stated in the

PGM's 22a regulation. In the new 22@bcn regulation, a clear mix of uses is considered, with some limits only on certain industrial and housing uses. Consequently, the following uses may coexist in the new sub-area.

a   *Industrial* — except for incompatible categories and uses relating to transport logistics. The coexistence of the different uses requires the definition of a compatibility framework from which harmful, polluting or dangerous industrial activities must necessarily be excluded.
b   *Offices* — permission for this use is necessary to allow for the introduction of new activities. To avoid the area's business activity becoming overly tertiary in nature, this use is only allowed for transformation purposes carried out via Special Plans.
c   *Housing* — under specific conditions:
    – housing in existing buildings is directly permitted without any planning procedure;
    – extension of existing buildings is permitted where this is presented jointly and when the buildings occupy a well-consolidated facade;
    – transformation operations must maintain a floorspace to land ratio of 0.3 of the overall plot and be protected (affordable) housing;
    – the re-use as housing of existing industrial buildings is permitted in buildings which do not exceed the floorspace/land ratio of 2.2 and which are included in the catalogue of listed buildings and meet the criteria established by the regulations.
d   *Commercial* — excluding hypermarkets.
e   *Residential* — hotels, flats, rented accommodation linked to workplaces, i.e. buildings linked directly to companies in the area which are used for temporary accommodation by personnel working there.
f   *Facilities* — those provided for by the PGM and compatible with it. In all cases, the needs of those living in the area will be met, in particular for health, sport and leisure facilities.
g   *A specific type of @ facility* — we may speak of facilities relating to @ activities and the specification of land allocated to this classification, where activities relating to training, research and enterprise may be carried out.
    These facilities comprise the specific allocation for the 22@ Area and must include research and publication of the R&D arising from collaboration between universities and business.

The field of research and training is increasingly linked to business and to the universities, to the extent that many of these activities are carried out in the same place. It is here that the land reserves proposed could play an important role.
@ activities are defined by their relation to the new information and communications technology industry, as well as those which, independently

of the economic sector to which they belong, are related to research, design, publishing, culture, multimedia and database and knowledge management. Included in an appendix to the regulations is a list of the activities embraced by the @ activity concept and an adaptation mechanism. The aim is to encourage the presence of these economic activities in the new area.

The identification of @ activities is rendered necessary by the fact that the Modified PGM provides for a greater intensity of plot development if the transformation operations include activities of this type.

Transformation management.

Barcelona City Council has created a firm, 22@bcn S.A., entirely funded by municipal capital, to serve as the main force behind the development of provisions contained in the Modification of the PGM. In this way, a legally independent management agency has been created, bringing together the necessary instruments and expertise for the management of the transformation process.

New infrastructures

A special Infrastructure Plan for the area represents an investment of over 27 thousand million pesetas, including the following elements.

1. *Energy* — the energy systems planned are those derived from renewable energy sources, in accordance with current environmental regulations, with direct electricity and gas network connections. Additionally, the construction is proposed, with the prior municipal concession, of an air conditioning network with centralized hot and cold water.
2. *Telecommunications* — all telecommunication operators currently in existence will have access to the area. Space is also reserved for future operators.
3. *Water cycle* — as regards the supply of drinking water and sewerage, the Plan proposes modernizing and completing the existing network. Elsewhere, the Plan proposes the rational use of existing phreatic or groundwater for compatible uses in Poblenou.
4. *Cleaning and selective collection of waste* — the Plan proposes the construction of a pneumatic waste collection network, incorporating elements and techniques enabling selective collection. Also proposed is the reinforcement of existing collection centres and the construction of new ones to ensure the balanced operation of the system.
5. *Public and private space* — the general principle of the Plan is that only transport infrastructures (trunk telecommunications networks and medium and low voltage electricity) should pass through public areas. The distribution network and the installation and equipment of the various networks (transformers, nodal centres, etc.) should always go

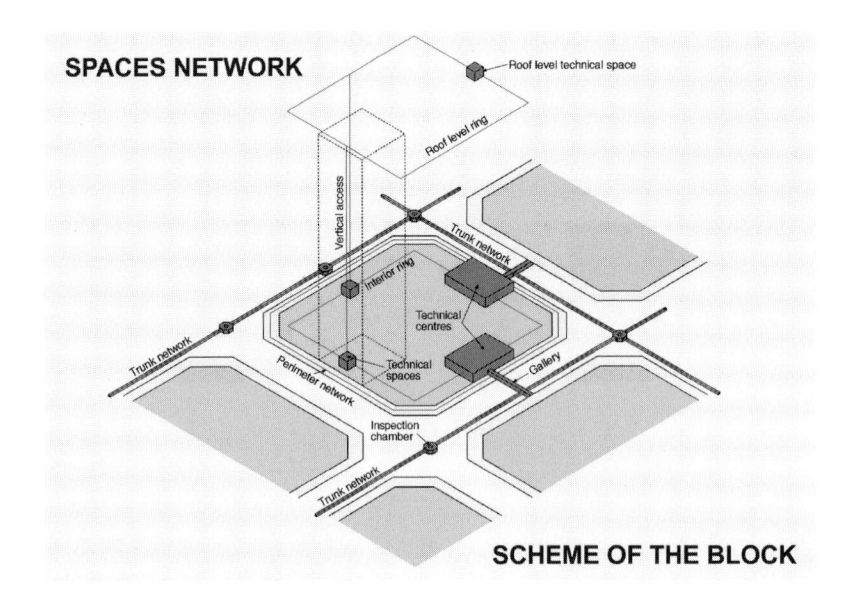

11.3
Axiometric view of infrastructure proposals for blocks in 22@ area

inside the privately managed areas. Similar treatment is given to the loading and unloading system, which will be in bays inside the blocks of buildings, as well as the waste collection points.

6 *Channelling systems* — the plan proposes a mixed system of service galleries and channels. Galleries run transverse to the streets and are aimed at connecting the blocks and providing space in which to work on the installations.

7 *Transport and mobility* — it is planned that, in ten years' time, 48% of people going to Poblenou should do so by public transport, 29% by private transport, and 23% by other means. With regard to road traffic, a hierarchical system is established for roads in Poblenou, differentiating between primary and secondary routes. With regard to traffic management, a centralized intelligent system is proposed to coordinate traffic lights and variable signalling. The parking system is planned as a mixed system managed by the private sector for parking for residents and employees, and a more flexible system for temporary and visitor parking.

The services for each island are centralized and distributed via the new gallery trunk network (Figure 11.3).

Proposed development planning instruments

The complexity of the transformation requires the definition of a flexible planning system to allow the following:

OPERATIONS DIRECTLY INITIATED BY THE CITY COUNCIL
These entail operations proposing the physical transformation of land for the creation of new structuring elements or for the introduction of activities which

may play a strategic role in the creation of new sector dynamics. Six areas of transformation have been defined: Llacuna, Campus Audiovisual, Parc Central, Pujades – Llull (Llevant), Pujades – Llull (Ponent), Pere IV – Paraguai. As they are concerned with operations that play an important part in defining the structure of the sector, it has been decided that the Special Plans for development of these areas will be brought about by public initiative. As a whole, these account for over 40% of the land under transformation.

OTHER TRANSFORMATION ACTIVITIES IN AREAS NOT YET DEFINED WHICH SHARE THE SAME AIMS AND CONTENT AND WHICH MAY BE DEVELOPED BY PRIVATE INITIATIVE

The rules of the Modification of the PGM govern the carrying out of transformation operations by means of Special Internal Reform Plans (PERIs):

1. the basic criterion is that the area of planning covers an Eixample block or a block defined by a passage;
2. the area of planning must be adjusted according to the existence of consolidated housing facades and industrial buildings that may be excluded;
3. based on this "corrected" area, it is established that the transformation action corresponds to 60% of this;
4. it is permitted that owners with a significant percentage (60%, or 80% in the case of blocks with passages) of this planning area may formulate the plan;
5. under no circumstances is public initiative excluded.

Conditions for transformation

1. Building land coefficients:*

Former coefficient (PGM 1976)	2
Present coefficient (MPGM 2000)	2.2
With addition of @ activities coefficient (+0.5)	2.7
With addition of public housing coefficient (+0.3)	3.0
With addition of complementary coefficient (+0.2)[†]	3.2

[*] Unit of building land: building land is measured in terms of square metres of floorspace that may be built per square metre of land in the plot.
[†] In priority transformation areas directly initiated by the City Council.

2. Standard land allowance: 31 m² for every 100 m² of housing (*18 m² per green area*). 10% of the area is aimed at facilities and the other corresponding allowances, which will be used preferentially for protected housing.
3. Special Infrastructure Plan Endowment: the Modification of the PGM expressly proposes that a part of the Plan's costs shall be met by the landowners, provided this land is not used for housing.

4   The Modification of the PGM maintains the aim of more precise adherence to the specific conditions of each action unit (by the derived plan (PERIs)). It expressly obliges that the Special Plans guarantee the fulfilment of the principles of community participation in gains generated by planning and of the balance between benefits and duties.

The extent of the transformation

The MPGM acts over an area of 198.26 hectares, broken down as follows:

1. the transformation of 1,159,626 m² of industrial land, with the potential for approximately 3,100,000 m² of new roof area, excluding facilities;
2. the official recognition of the 4,614 residences currently on industrial land;
3. approximately 2,700,000 m² of floorspace for new economy activities;
4. approximately 4,000 new residences, under official protection;
5. an increase in green zones of approximately 75,000 m²;
6. the creation of approximately 145,000 m² of land for new facilities;
7. an increase of approximately 60,000 jobs.

Appendix

List of @ type activities

ICT

- Manufacture of computers and ancillary equipment
- Manufacture of computer consumables
- Manufacture of telecommunications systems (telephone exchanges, network control systems, cell phone systems, satellite communication systems, etc.)
- Manufacture of telecommunications equipment (terminals, etc.)
- Manufacture of telecommunications cables (copper, optical fibre)
- Manufacture of electronic items including television, radio and communications equipment
- Playback systems for recorded media/data
- Development, production, supply and documentation of computing programs
- Production of management and control software for smart management of telecommunication networks
- Radio broadcasting and telecommunications
- Development of transmission by cable
- Internet
- Multimedia
- Paper-based publishing (newspapers and magazines)
- Audio-visuals

SERVICES

- Data processing
- Email-related activities
- Database-related activities

- Providing added value services (email, data exchange, EDI electronic fund transfer, EFT video conferencing)
- Supply of digital goods and services
- Computer maintenance and repair: provision of technical services, hotlines, support, maintenance, outsourcing, after-sale services
- Other telecommunication services: all activities related to cell phones, satellite communications and their applications to other sectors such as transport and distribution
- Services for the creation of new companies
- Services for improving the competitiveness of existing companies:
    – technology
    – sales-related
    – finance
    – administrative
    – personnel

CENTRES OF KNOWLEDGE

- Further education centres (professional training, schools, etc.)
- Universities and centres of continued education
- Research centres (public and private R&D)
- Cultural amenities (museums, libraries, etc.)
- Professional associations
- Information, documentation and advice centres
- Publishers and companies involved in audio-visual creation
- Knowledge-based companies
- Artistic activities and those involving management in the culture field

Note

The complete text of the MPGM and several Special Plans may be found at: www.bcn.es/22@bcn

PART 3

Critical perspectives

The final chapters represent only a small portion of those debates and writing within Barcelona which do not see the changes as an unalloyed success. Probably no one is arguing that there has not been an enormous amount of good work since the 1970s. That is the solid bedrock of support for a more attractive and civilized city and way of life, as that has been built up over these years. However, there are voices, mainly from the left and green parts of the political spectrum, which, whilst going along with much of the achievement, take a significantly critical stance. Two are introduced here. The first, by Mari Paz Balibrea, an anthropologist now teaching in London, presents the case for a quite different analysis of the political and cultural signification of the changes. She implicitly raises the argument that much could have been done very differently, to the long-term advantage of more people, creating a more liveable and less "uniform-modern" urban reality.

Enric Tello presents a reflection on where Barcelona fits into the environmental sustainability discourses and realities of the past ten years or more. He has been an active participant in this matter, pressing the council to take green issues far more seriously — with some achievements to the credit of this pressure, though with massive challenges unmet.

Chapter 12

Urbanism, culture and the post-industrial city: challenging the "Barcelona Model"

Mari Paz Balibrea

In 1999, precedent has been broken to award the Royal Gold Medal to a city: to Barcelona, its government, its citizens and design professionals of all sorts. Inspired city leadership, pursuing an ambitious yet pragmatic urban strategy and the highest design standards, has transformed the city's public realm, immensely expanded its amenities and regenerated its economy, providing pride in its inhabitants and delight in its visitors [...] Probably nowhere else in the world are there so many recent examples, in large cities and small towns, of a benign and appropriate attitude towards creating a civic setting for the next century.[1]

The above quotation comes from a press release by the prestigious Royal Institute of British Architects (RIBA) upon the award of their Gold Medal to the city of Barcelona in 1999. This text is paradigmatic of the dominant perception of the city as seen from abroad: a stylish and exciting metropolis, the perfect civic site for the urban communities of the twenty-first century. The "Barcelona model" has gained official approval and is being copied on an international scale. Still in a British context, it is well-known that Barcelona is considered to be a privileged urban reference point by Tony Blair's cabinet. In 1999, the Urban Task Force created by the Labour government and led by Lord Richard Rogers produced a plan for the regeneration of ten UK cities based on the urban programme of Barcelona. *The Observer* summarized the aim of the project as follows: "Each of the target cities will be encouraged to sell themselves as exciting and stylish places to live and work, mirroring the success of the Catalan urban regeneration" (Wintour and Thorpe 1999).[2]

The view from inside is not much different. Whilst in the rest of Spain social democracy has been showered with criticisms as a result of the corruption and speculation condoned and stimulated during the Socialist Party's period in office, resulting in their being ousted from government in 1996, the majority of people in Barcelona remain satisfied with the way the local Socialist government has managed things, particularly with respect to its urban and architectural projects, and continue to vote for them in local elections. Despite the economic recession that followed the 1992 Olympic Games, Barcelona has continued to enjoy an uninterrupted heyday of national and international prestige, as well as practically unanimous consensus with regard to the quality and beauty of its urban developments and the habitability of a city seen as Mediterranean as well as "human".

Of course, the absence of any notable dissent among the city's people has been brandished as irrefutable proof of the virtue of the urban changes which have been implemented (RIBA's opinion is just one of many examples). For others, however, this consensus is a sign of "citizens internalizing criteria which coincide with the interests of the dominant economic interests" and they therefore consider it to be "one of the most serious aspects of recent political and social processes" (Etxezarreta et al. 1996: 289). From an ideological point of view, the production of consensus is the principal means of legitimizing domination and of co-opting potentially critical citizens (Ripalda 1999: 30; Esquirol 1998: 113–30). Any hegemonic ideology will seek to devise for its "interpellated" (i.e. locate, construct) subjects a representation of reality that, while favouring its own interests, can at the same time be presented as the only truth about that reality. This chapter correspondingly assumes that the popular consensus on Barcelona needs to be regarded with scepticism and vigilance, particularly in view of increasing social polarization, the growth of a peripheral population which has seen its quality of life deteriorate since the 1980s, and the massive speculation accompanying the restructuring of the city (Roca 1994). In what follows I will analyse the urban changes that, since the early 1980s, have produced the seductive Barcelona of the 1990s. I will additionally define some of the major mechanisms through which the perception of those changes has been constructed, paying particular attention to the role played by culture. My aim is to expose the ideological and political underpinnings sustaining the consensus described above. This chapter, therefore, focuses on the dominant and institutionalized forms in which the process of generating social space and deriving meaning from it has taken place in post-modern Barcelona. I am aware of the importance of spatial, urban practices and experiences that are initiated as contestations to the dominant consensual perception of the city, but will not deal with them here.

The city as ideological text

As soon as we think of urban spaces as texts, and therefore as vehicles of ideology,[3] then urbanism and the production of consensus become interconnected processes. Urban and architectonic built spaces constitute privileged sites within which ideological interpellation takes place. To give shape to the collective sphere through an urban regeneration project is also to semanticize (or resemanticize) the former; like every signification process, this is intensely ideological (Ramírez 1992: 173–82). It becomes crucial to know what is being built in the city and how the newly built spaces are endowed with hegemonic meaning, in order to understand how individuals and collectives are ideologically interpellated as citizens. As Georg Simmel argued long ago "The production of spatio-temporalities is both a constitutive and fundamental moment to the social process in general as well as fundamental to the establishment of values" (quoted in Harvey 1996: 246). Fredric Jameson specifies:

> the building interpellates me — it proposes an identity for me, an identity that can make me uncomfortable or on the contrary obscenely complacent, that can push me into revolt or acceptance of my antisociality and criminality or on the other hand into subalternity and humility, into the obedience of a servant or a lower-class citizen. More than that, it interpellates my body or interpellates me by way of the body [...] (1997: 129).

As soon as they come into being, buildings and urban spaces signify. First of all, this is because they change the structure of perception in the everyday urban experience of citizens. Let us take, for example, the recent opening of new avenues and arteries in Barcelona, such as the Rambla del Raval, the carrer Marina and the extension of the Avinguda Diagonal and carrer Aragó to the sea. In the old city, the Casc Antic, the longitudinal demolition of entire blocks of houses was necessary to make room for the Rambla del Raval; in the north-eastern areas of Poblenou and Besòs, which have been completely redeveloped and restructured, all changes have followed the closing down of the local industries and the revaluation of the land they once occupied. Such spatial changes can generate positive effects for the citizen, such as a new sense of cleanliness and rationalization producing pride and satisfaction with the current configuration of the city; or negative effects such as a sense of alienation and displacement at the loss of the original habitat (Terdiman 1993: 106–47; Benjamin 1973). The result will depend on the citizen's previous relationship to the now transformed spaces and on the material and symbolic conditions under which she has experienced the change, and will also be affected by the degree of persuasiveness of the different discourses circulating and giving meaning to the changes. In the case of Barcelona, these discourses have overwhelmingly, almost monolithically, been favourable to the urban changes implemented in the city.[4]

Citizens are not the only targets interpellated in the process of resignifying the city. In accordance with the logic of the tourist industry (as we shall see, a fundamental feature of the city's current economy), the entire city turns into a lucrative, luxury, fun commodity that can be rapidly consumed by the tourist, a leisure space commodified repeatedly in the purchase of a plane ticket, a book on Gaudí, tickets for the opera at the Liceu or for a concert at the Palau de la Música, the booking of a hotel room or restaurant table. In each and every one of these activities, all of them marked by an economic transaction, the hypothetical tourist "buys" the city and constructs a private imaginary of it: one that, to a great extent, is previously manufactured for her by multiple local and global practices and interests. Even — or especially — in the case of those activities not involving an immediate act of consumption (for example, the following of recommended tourist routes such as la ruta del Modernisme) the semantics and hermeneutics of space have been constructed for the foreign viewer, and this construction has necessitated a previous political and economic intervention in the form of the restoration, face-lifting and rehabilitation of buildings, the equipment and staffing of venues, the production of targeted bibliographies, etc. Dean MacCannell argues that the therapeutic quality of the tourist trip stems from the tourist's desire to create a totality out of the visited space, one that saves her from the everyday fragmented reality surrounding her in the modern world (1976: 7, 13, 15). Such a totality can be obtained in an alien environment because the tourist can reduce this new reality to a very limited number of experiences, and its past to a few visits to museums. Barcelona as a leisure and tourist site needs constantly to produce a totalizing and coherent representation/meaning of the city, one that is easy and pleasant to consume for this kind of visitor.

Not unrelated to this logic of resignification, sometimes the construction of a new space responds to an urban reconfiguration in need of new privileged signifiers that can be used to represent the redefined city as a whole. This is the case of architectural and urban projects such as the Foster telecommunications tower on Tibidabo or the Port Olímpic where a new leisure area is located. Due to the social function performed by these new spaces and artifacts, they become the signifiers best suited to symbolize or synthesize the current dominant meaning of the city as a Mediterranean centre for leisure, communications and high-tech industry. Today, these places figure on all tourist routes and their icons appear on many tourist maps as well as on every visual representation of Barcelona financed by the City Council, for instance in the paradigmatic work of Javier Mariscal.

The prominence acquired by these new signifiers in recent years works against the symbolic status previously enjoyed by other urban and architectonic spaces of the city, notably those which allude to its industrial past. Most of the land that has been turned into service, leisure or residential areas since the 1980s has been produced by the revaluation of old industrial land, and the demolition of the old factory buildings previously occupying it. As these disappear en masse, some have been salvaged and given new functions in their entirety (often as cultural, sometimes as residential spaces), while desemanticized fragments of others have been preserved as monuments: for example, the chimneys in the Poblenou district (from an old textile factory) or in the Parallel district (from a former power station). These fragments not only lose any practical function, but in their new location their socially symbolic potential is also reduced. In theory, such fragments (for example, a chimney formerly used to extract fumes) have the potential to become symbols of bygone socio-economic activity, an allusion to the city's past. Or do they? These now monumentalized "objects" undoubtedly refer to the past, but their spatial recontextualization, the new syntax of space, disconnects them from the local history in which they originated. Isolated in the middle of areas now reconverted into shopping malls, new residential complexes for the middle classes or luxury offices for business executives, they can hardly be more than flat and mute citations. And in so being, they are rendered increasingly unable to convey a sense of their own historicity to those ignorant of local history. The historicity of these places can be deciphered through historical knowledge, but much less through the experience of bodily interaction with the new spatial configurations. Indeed, these fragments' disposition in space conceals the complexity of an industrial past characterized by social struggles and human relationships that were lived out on that spot, replacing it with a new configuration of space which promises the absence of conflicts and equality through consumption and the market.[5] In other words, they fail to capture a culture, and a politics, of the place where they come from. Much the same could be argued in the case of those old industrial buildings given new functions as cultural spaces. The potential allusion to the past that their mere presence invokes has been restored, aestheticized, to the extent that the end product mostly loses its capacity to refer to a critical and complex memory of industrial/capitalist development and exploitation and of the role that this complex history, imbedded in the building and its surroundings, has played in the city's current prosperity. In those areas where there has been a strong process of gentrification, these architectural quotations are even more alienating, in the sense that they are no longer connecting with the memory of those neighbours who had dwelled there in the past. Those who had used the spaces, and their families, are no longer there to turn them into everyday spaces of shared memory.

Under such conditions, the architectural quotation of the past[6] paradoxically promotes amnesia and an absence of reflection on history. This new monumentality turns the object from the past into an empty shell, a liberated signifier with a highly tenuous and malleable signified attached to it, whose quasi-floating character can be easily appropriated both to justify the local government's interest in keeping the memory of the city alive, and to serve as the logo of a shopping mall. Thus, its resignification is, at the same time, a desemanticization. Those who favour this desemanticized conservation-production-resignification of space participate in and benefit from the politics of history implied in such uses of these objects from the past.

Finally, there is also a political/ideological dimension to aesthetics that is relevant to the urban text. Architecture and urbanism are applied arts implemented only when and if there is sufficient capital to finance their projects and sufficient political power to back them. Urban speculation, via the large-scale real-estate business, is a classic phenomenon in the history of local corruption sustaining every urban development process. But the production of form and beauty (the terrain of the aesthetic) is also an intrinsic component of these disciplines. The presence of a mutual tension and interchange between aesthetic and politico-economic considerations can never be avoided in the fields of urbanism and architecture. What is peculiar to democratic Barcelona with respect to this tension is the resignifying of space, which has in turn produced an intensification of the aesthetic component. One example comes immediately to mind: between 1986 and 1999 the city council spent 6,923 million pesetas on its campaign *Barcelona posa't guapa*, intended to promote and subsidize the face-lifting of key buildings located around the city. According to ex-Mayor Pasqual Maragall, the embellishment programme promoted through the *Barcelona posa't guapa* campaign "consolidates the citizen's perception of public space as a common good, contributes to the improvement of the collective heritage and increases the comfort, tranquillity and sociableness of the city" (Ajuntament 1992: 6). It should also be added that this campaign has massively benefited the modernist architectural patrimony of the city, mostly located in the Eixample district (Ajuntament 1992: 1999). It is not coincidental that this modernist heritage, dominated by the work of Antoni Gaudí, has become one of the pillars sustaining the city's tourist cultural provision and, more generally, of its constructed image and personality. The proliferation of similar restorations, together with a policy of awarding the category of listed building to some of these restored properties (which means that they are considered part of the city's, that is, part of a public, patrimony), has become an enormously successful political strategy, which brilliantly exploits the one communal, collective aspect of most of these buildings: the fact that their facades can be seen from the public space of the streets.

The entire process of urban transformation in Barcelona is carried out under the technical supervision of a group of notable architects and urbanists. To understand the quality and attractiveness of many of the urban projects that nowadays constitute Barcelona's urban space one has to take into account this group's achievements in the pre-Olympic period. But the exceptional media attention that this group has attracted is also a crucial factor in explaining why this quality and attractiveness have become a universal focus of attention, popularizing the dictum that Barcelona is "the city of architects" (Moix 1994a). The quantity and quality of the work of these professionals has aroused much admiration for the city, to the point of becoming the only, or at least the privileged, element by which it is judged and thus a decisive component of the city's seductive appeal. Until the mid-1990s, the rehabilitation and face-lifting of entire key areas created new urban spaces and cultural facilities, many of them public, which without fail include an architectural project by a named architect. The presence, and convenient marketing, of work in the city by Foster, Meier, Viaplana, Calatrava, Isozaki, Moneo, Miralles and others intensifies the aesthetic signification of these projects, turning them, by the same token, into privileged signifiers of what I have previously defined as a "designer" city.

The constant tributes paid to the city's beauty have helped to distract the attention of visitors and citizens alike from other fundamental, much less satisfactory, issues: employment, housing, public transport, or even the questioning of the same urban projects whose aesthetic value is so intensely praised. One could say, provocatively drawing on Walter Benjamin's famous dictum (1969: 217–51), that the more aesthetics is politically used in Barcelona, the more politics is itself aestheticized, so that political consensus and the obedience of the masses are achieved by continually producing for them what is perceived as aesthetic or artistic gratification. It is clear, then, that a city is an ideological text. Let us move now to a more systematic analysis of how this text has been (re)written in key modes, how its forms have been filled with meaning, and how spatial changes have helped to shape a political ideology in democratic Barcelona.

Culture, urbanism and the production of ideology

In my opinion, the most remarkable phenomenon in the process of urban change, that is, in the major transformations undergone by Barcelona's urban fabric/text, is the extraordinary broadening of the material and symbolic terrain occupied by culture. The most important changes affecting the social body and the economy have been justified in the name of culture, which becomes their structural axis. By invoking culture, the ideological continuity of the consensus with regard to the city has been made possible. The connection between culture and urbanism is established through the latter's capacity to create public space: a public space rhetorically defined as open to all,[7] and therefore as the place of encounter and of the production of collective culture. The new urbanism in Barcelona, at least up until the first half of the 1990s, is characterized by the enormous proliferation of cultural spaces (that is, spaces that house "Culture" with a capital "C") and public spaces of great impact and visibility: squares and monuments came first, then museums, theatres, sports complexes, avenues, promenades. The beginning of this trend dates back to the period of the Transición (Gomà 1997; Etxezarreta et al. 1996; Bohigas 1985) and the early stages of Socialist leadership of the council by the mayor Narcís Serra and then Pasqual Maragall. The discourse on the need to monumentalize and rehabilitate the city so as to serve its citizens, creating spaces of identification for the community, and the pressure exercised on political leaders by important grassroots residents movements in order to make things move in this direction, are paradigmatic trends of this period.

As Oriol Bohigas, the head of the City Council's urban planning department, put it, the task was to "monumentalize" the outskirts, "giving them significant collective value", and to restore the historic centre by reversing the process of decline it had suffered, thus helping to "improve collective consciousness" (Bohigas 1985: 20). The aim of this urban policy was to produce a greater uniformity in the city, in the sense that the residents' cultural capital and collective memory would be recognized within each neighbourhood. The homogenization which Bohigas aimed to achieve by intervening in the most socially deprived areas of the city was, in reality, an attempt to bridge glaring inequalities and to foster social reconciliation by allocating cultural and symbolic capital in the form of (among other things) monuments and public spaces of socialization and memory.[8]

The continuities and the ruptures in the discourse and implementation of urban transformations in Barcelona throughout the democratic period must be understood, at least partially, in relation to this earlier stage. The major urban projects that would from the mid-1980s characterize Barcelona's urbanism were largely justified through a rhetoric which promoted construction as a public service designed to benefit the everyday life of citizens, including those most disenfranchised. The logic and politics behind these earlier projects would get lost from view with the exponential escalation, in terms of quantity and quality, of the urban projects that followed. What triggered the change? The extent of the social and economic transformations undergone by the city was a result of the economic restructuring of the entire Spanish state that was being effected from the 1970s onwards. In the case of Barcelona, the transformations consisted in the dismantling of much manufacturing industry, which turned Barcelona into yet another example of what has been come to be known globally as a post-industrial city. Under these circumstances, it became imperative to look for a way to make the local economy sustainable. This signalled the beginning of a shift in the city's economy towards "clean" industries devoted to the production of culture and technology:[9] a local manifestation of a worldwide process where information and entertainment become the driving forces, necessitating the proliferation of sports infrastructures and large commercial and cultural centres (Cohen 1998: 110; Harvey 1996: 246; Smith 1996). For those with major economic interests in the city, the urban transformation required by the projects for monumentalizing the outskirts and for regenerating the historic centre came at just the right time, allowing the economic recession to be tackled. It was a first step towards the reconfiguration of the city, and the same economic sectors would exercise pressure in order to make the transformations more and more dramatic. (Etxezarreta *et al*. 1996; Gomà 1997). The pressure exercised by big financial groups over local governments or, eventually, the alliance between the two, translated materially into the promotion of a mixed economy as the main way of financing public urban redevelopment. Politicians presented this as the domestication of capital through its harnessing to the priorities of the welfare state (Bohigas 1985:

63), or simply as an inevitable requirement in order to finance the public spending necessary to transform the city.[10] From the point of view of private interests this was a very beneficial, if not indispensable, alliance. The creation and improvement of the city's urban facilities and spaces would allow the recapitalization of their industrial land, devalued after the economic recession and now very lucratively revalued and recycled. This would help to promote the cultural, technological and entertainment industries that were to become the core of the new economy (Harvey 1989: 260). Pasqual Maragall expressed this very clearly in 1991: "in urban competition factors like the environment and cultural and educational infrastructures count more and more. In a strategic sense we can say that cities are like businesses which compete to attract investments and residents, selling places which are suitable for industry, commerce and all kinds of services" (1991: 99). The single most instrumental event triggering, encouraging and justifying all of these "necessary" changes was the nomination of Barcelona as a candidate to host the Olympic Games. I will come back to this.[11]

At this precise historical conjuncture — the beginning of the 1980s — the interests of grassroots residents movements coincided with those of business and financial groups and those of local government. All of them, for different reasons, were in need of more public space and improvements to the quality of urban life.[12] But the huge spatial redefinitions that were to be implemented in the city ended up negating the original planning principles. The small-scale, detailed, respectful urban project intended to bring direct benefit to the most depressed neighbourhoods and their inhabitants had very little in common with the huge structural transformations required to implement projects such as those of the Vila Olímpica, Poblenou, Diagonal Mar or Sant Andreu/Sagrera, all of them built in what had previously been industrial and/or working-class areas. Even though the majority of these "areas of new centrality" (Ajuntament 1991), labelled as priority targets for urban development, were located in very run down, peripheral zones, the new developments were not aimed at benefiting the existing local population. The extremely lucrative revaluation of the old industrial areas, located next to marginal, working-class neighbourhoods, where most of the new cultural, sports, entertainment, commercial and residential facilities were built would end up producing precisely what the urban planning policy of Oriol Bohigas had wanted to avoid:[13] the dislocation of the original neighbourhoods. The proliferation of big shopping malls has accelerated the destruction of the social fabric of small local businesses (Grup d'Estudis Territorials i Urbans 1999) and impoverished the quality and quantity of public spaces, in favour of pseudo-public (but in fact restricted and private) consumer locales. Many inhabitants of these neighbourhoods or their children have de facto been expelled from their historic communities, unable to afford the escalating prices of new residences in their now improved areas, or forced out of buildings expropriated for demolition.[14] By working ideologically as a rhetorical instrument for generating consensus and consent on the part of the population, the process of monumentalizing the outskirts and of improving public spaces around the city has, paradoxically, facilitated the transition to a situation of progressive gentrification, privatization and more and more restricted access to public spaces.

In this new situation, we are witnessing the progressive erosion of the meaning of the term "public" and, with it, the redefinition of the space occupied by culture. This is not only because culture is more and more radically commodified and dependent on private producers deciding who has access to their (also private) spaces of consumption (Rifkin 2000). Culture is also redefined because those representing the public interest (local governments) now understand culture as a key industry for the local economy, and not just as the symbolic realm where ideology is produced or as the realm of aesthetics. According to Pep Subiròs, Head of the Olimpiada Cultural and also then director of the Centre de Cultura Contemporània (CCCB):

> In the irreversible reconfiguration of large cities as service centres, culture plays a basic role. This is not just a matter of having a good supply of events, prestigious museums, for internal and tourist consumption. You also have to, perhaps most importantly, have the capacity to receive, recycle and export ideas, sensibilities, projects which improve the internal quality of life and upgrade the city in international competition. No city with a rich cultural life lacks solid cultural structures and resources for contemporary creativity (1989: 6).

According to this interpretation, culture ("ideas, sensibilities, projects") is a commodity that needs to be produced in a competitive market, for which purpose certain means of production ("cultural structures and resources") are also necessary. Pasqual Maragall himself published *Refent Barcelona* (*Remaking Barcelona*) in 1986: a key book for understanding his government's political vision as well as its specific objective of transforming Barcelona's economy by gearing it towards the tourist, technology and cultural industries. He states that Barcelona has the capacity to become the northern capital of the European south,[15] given its economic strength and its very attractive Mediterranean "art of living": "Beyond the existing reality of the Catalan lands, we have to go further to become the 'European north of the south' and move in the world market of culture, tourism, economic investments and so on. The possibilities are enormous" (1988: 95).

This incorporation of culture into the economic realm becomes an even more complex issue when considered alongside that of Catalan nationalism at a time when the Spanish state was democratizing and starting to recognize the rights of the historic nationalities comprising it. Within this propitious climate, Barcelona has been able to consolidate itself, politically and symbolically, as the capital of a Catalan nation without a state, in accordance with all modern Western nationalist projects. Cultural facilities, those housing

Culture with a capital "C", have played a key role here. The conception of the Museu Nacional d'Art de Catalunya, the reconstruction and extension of the Liceu, the building of a Teatre Nacional de Catalunya, of a new Arxiu de la Corona d'Aragó, or of the Auditori, have been implemented by the autonomous and/or local governments as ideological instruments in the construction of a nineteenth-century-style nationalism. More specifically, these centres were conceived of as spaces which would organize culture from above and stockpile the cultural capital of the local elite while, at the same time, educating the masses in appreciation of this cultural heritage's supposed public and community value (Duncan 1991). Other Spanish cities such as Madrid (Centro de Arte Reina Sofía), Bilbao (Museo Guggenheim) and Valencia (Institut Valencià d' Art Modern), have had a policy of embodying their "state" politics in one large cultural centre by a big-name architect, the container being as important as its contents. Barcelona, however, has chosen to disseminate the politico-symbolic meaning projected by the cultural space in multiple architectural interventions. Of course, as in the other cities mentioned (the Guggenheim in Bilbao is perhaps the most obvious example), these centres, apart from consolidating the nation's symbolic capital, are integral parts of the area's tourist appeal (Walsh 1995) and function as spaces where the national heritage can be consumed. They are, as Neil Harris (1990) provocatively calls them "department stores of culture".

This superposition of modern nationalism onto a "post-industrial" economy and a post-modern cultural logic does not always produce the harmonious convergence of interests we have mentioned above. While Catalan nationalist discourse has supported every space which symbolized the intensification of national identity and collective memory, in a global context in which the nation-state is being weakened and questioned, not all projects in the city have served to consolidate that particular nationalism. The long-standing rivalry between the local Socialist government of Barcelona (PSC) and the autonomous Catalan nationalist government (CiU) over who has most successfully appropriated the meaning of Barcelona, and their disputes over the awarding of cultural funding, reveal two different responses to nationalism in post-modernity. I have referred earlier to Pasqual Maragall's projected definition of Barcelona as the "northern capital of the European south". Within the same paragraph he warns against the danger of Barcelona becoming associated with European nationalities "smaller than Catalonia" or forming part of an "international of oppressed nationalities." By articulating these "dangers", Maragall seems to be dissociating himself from what might be perceived as a form of nationalist victimization or essentialism, making clear that nationalist[16] demands are not top priority in his political agenda, even though at some points he invokes Catalunya and the Països Catalans in his framing of the question. This is not at all surprising coming from the local leader of a Spanish political party, the PSOE; but it seems to me that his anti-nationalist stand here does not obey a centralist logic conceiving of Spain as a unity with a radial centre in Madrid.[17] Rather, I interpret it as a positioning with respect to a new historical conjuncture, in which the local urban unit is called upon to have a new prominence, one not necessarily subordinate to any national framework (be it Catalan or Spanish) and therefore detached from its secular, political claims. In other words, Maragall is appealing to a local-urban rhetoric that might be called "nationalist" only to the extent that it promotes a discursive interpretation of social, cultural and political practices as characteristic and reinforcing constituents of collective identity, which determine whether social consensus is reached or not. In that sense, the social-democratic local government has promoted a "nationalism" for the city-nation with its own organization and imaginary community, but one that ultimately aspires (and this is the key differentiating element with respect to earlier or current forms of local patriotism[18]) to become part of a broader community of cities not marked or limited by state and/or national borders. While it is not clear from our historical perspective that this form of imaginary community and economic framework are going to become dominant, we can certainly call it a response to a new post-modern configuration of globalization where, as the Royal Institute of British Architects reminds us in the above mentioned text: "cities as much as nations are in direct competition for jobs and investment".

Cultural facilities such as the Museu d'Art Contemporani de Barcelona (MACBA) or the Centre de Cultura Contemporània de Barcelona (CCCB), regardless of their very different and contentious cultural politics, interpellate community members as citizens of Barcelona, and not necessarily (though the two are not incompatible) as Catalan or Spanish. The CCCB is financed exclusively by the Diputació and the Ajuntament de Barcelona, and not by the Generalitat de Catalunya or the central government, and it receives a small amount of private funding. According to Josep Ramoneda, the Centre's director since 1989 and very close to PSC political positions, the CCCB "is a project led by an idea, of urbanity as a real tradition of the modern world" (1989: 4). And elsewhere: "the concept defining our profile is the city, as a place in which modernity develops. I think the moment has come for a critical review of modernity, and if this review seeks a form, the city gives it one" (Moix 1994b). The connection made here by Ramoneda between modernity and the city is a very familiar and widely accepted one. But it is at the same time a way of emphasising *one* aspect of modernity (that of serving as a driving force for urbanization), to the detriment of another fundamental creature of modernity: nationalism and the emergence of the nation-state. Ramoneda's emphasis favours a perception of the city as a unitary and independent unit, rather than as part of another entity, and therefore underplays the notion of Barcelona as capital of the Catalan nation without a state. In so doing, he is distancing himself and the Centre's cultural politics and ideology from a nationalist position.

On the other hand, regardless of how much Ramoneda insists on remaining at all times within a modern frame of reference, the CCCB project fits an altogether different historical context. Together with the MACBA, the

CCCB is a key urban project in the transformation of the city, located right in the centre of the "restored" Casc Antic and heavily implicated in the tertiarization and gentrification of the city to which we have referred earlier. Moreover, the fact that the CCCB is recognized to be an imitation of the Pompidou Centre in the Marais district of Paris — the "Beaubourg of Barcelona" — indicates a desire on the part of the Socialist local government to imitate their Parisian counterparts, not only by converting the Centre into their cultural flagship, but also by striving to make the Catalan city into a part of the tourist circuit of great cultural cities (cultural capitals) of Europe.

The MACBA is another, albeit very different, case in point. Under the direction of Manuel Borja-Villel, also its chief curator, it has also practised a politics of recognizing, assuming and defining its identity through its location in Barcelona (and not Catalunya or Spain), but in a resistant and provocative manner. Unlike the CCCB and its institutional policy of raising Barcelona's status to that of the great European cities, the MACBA has promoted Barcelona's peripheral, subordinate position in art circles as an opportunity to reflect upon modern art and to rewrite its history and its present from the margins (Frisach 2000). Be that as it may, the spectacular Richard Meier building which contains the museum is the closest thing in Barcelona to the Guggenheim phenomenon in Bilbao: a space conceived as a tourist attraction in itself.

We could not finish this section without mentioning the event that has been most instrumental in manufacturing consensus between the different, opposing conceptualizations and uses of culture and public space analysed so far: the Olympic Games (Figure 12.1). Barcelona's "Olympic fever", understood as the city's collective pride at having been chosen to host the Olympic Games, had been brewing since the city's selection for this role in October 1986, and to a lesser extent since its nomination as a candidate in 1983. A sporting, cultural and ideological event all in one, the Games succeeded totally in generating local patriotism and consensus, as well as in introducing the city to the world at large. The Games were construed as a project by all and for all, an event in which everybody could participate, and from which everyone would benefit in the form of municipal self-esteem. The planning and implementation of the Games constituted the crucial period for securing the city's economic future and, as stated earlier, marked a qualitative shift towards major restructuring projects. On an ideological level, the Olympics sutured the gap that separated the minimalist urban policy of the first social-democratic government from the maximalist, populist policy of Maragall and his team. Invoking the Olympic Games as a pretext, streets were opened, ring roads were built, hotels went up, cultural and sports facilities proliferated (Figures 12.2 and 12.3). The pretext was so effective because it appealed to local patriotism, generating in the process a collective desire to rise to the occasion (Figure 12.4). The Olympic Games generated a civil fraternity, materially embodied and reinforced in every architectural and urban built environment, which was

12.1
Foundation laying ceremony of Vila Olímpica, with Pasqual Maragall on left and Joan Antoni Samaranch in centre

12.2
Vila Olímpica, with the only industrial chimney retained in the district

12.3
Flats built around industrial water tower, in "five blocks" scheme

12.4
Barcelona publicity 1990s

perceived as required by the event. Once again, this was not a unique phenomenon. In the post-industrial city, spaces of cultural and sports consumption become, in the words of Harvey, symbols "of the supposed unity of a class-divided and racially segregated city. Professional sports activities and events like the Los Angeles Olympic Games perform a similar function in an otherwise fragmented urban society" (1989: 271). More than any other event, the Olympics contributed to the construction of a collective imaginary of democratic Barcelona for its citizens. Indeed, they made the inhabitants of Barcelona into citizens and, for that reason, they constitute a fundamental patriotic event that even produced its own foot-soldiers (Espada 1995: 138) in the form of "Olympic volunteers", the citizens who helped as marshals, etc. during the Games.

The end of the Games meant the end of their vast potential for promoting the city. To replace them, another macro-project has, since 1997, been proposed by the local government: that of "Barcelona 2004: Fòrum Universal de les Cultures". While the Olympic Games provided a great opportunity to redevelop and restructure the city, the Fòrum is designed to continue such a task, through a project where the centrality of culture in a trans-national context is even more obvious. The event will span the period between 9 May (reduced from the original 23 April, dia de Sant Jordi, Catalunya's patron saint and also the Day of the Book) and 26 September (dia de la Mercè, the patron saint of Barcelona, and the occasion for the city's "fiesta mayor", a popular cultural event). In one of the official brochures issued at an early stage, the Fòrum project is defined as a:

> call to all world cultures to make their voices heard, in a great shared festival ... it tries to respond to the challenge of globalization of human activities, from the action of civil society ... to bring together everything favouring solidarity, human rights and sustainable development.[19]

The use of terms such as dialogue, solidarity, human rights, civil society and sustainable development play a prominent role in defining a fashionable kind of progressive rhetoric which is in fact refuted by the increasing inequalities that the facilitation of this project is generating in the city.[20] The use of culture as an umbrella word or concept (Fòrum Universal de les Cultures), covering all of the areas mentioned, is indicative in itself of the leading role and totalizing meaning that culture is acquiring today. Joan Clos, current Socialist mayor of Barcelona, defends its promotion as a thematic replacement in a new context for the World Fairs and Universal Exhibitions characteristic of the twentieth century: "The century's world fairs have been technology supermarkets for gazers and consumers, but in the next century we want participative forums where ideas are debated on peace, sustainability and culture" (Altaió 2000: 237). What has not changed with respect to these modern events is their use by local governments and financial interests as ways of promoting the city and as occasions for restructuring and replanning it (Figure 12.5). In the case of the Fòrum 2004 project, a civilized, humanist proposal for dialogue among cultures, one that is respectful of difference, goes hand in hand with the urban redevelopment of the Besòs estuary area, one of the most depressed neighbourhoods but also one of the most coveted sites in the city, with an estimated investment of 170,000 million pesetas (Ricart and Aroca 1999; Cia 1999). Culture and urban redevelopment, culture and spatial restructuring continue to go hand in hand in Barcelona, continuing to justify each other.

In this chapter, I have focused on a critical analysis of the urban and cultural transformations defining Barcelona in the democratic period, paying particular attention to the discontinuities and changes that, at the level of hegemonic discourses, are presented as consensual continuities. My contention has been that the profound spatial changes taking place in democratic Barcelona obey the logic of a progressive "tertiarization" and the post-industrialization of the city's economy, a process bringing with it the imposition of a certain post-modern hegemony.

12.5
Construction of Diagonal Mar park and flats, c. 2000

References

Ajuntament de Barcelona (1991) *Àrees de nova centralitat*, Barcelona: Ajuntament de Barcelona.
Ajuntament de Barcelona (1992) *Barcelona, posa't guapa. Memòria d'una campanya*, Barcelona: Institut Municipal del Paisatge Urbà i la Qualitat de Vida.

Ajuntament de Barcelona (1999) *Barcelona, posa't guapa. 13 anys*, Barcelona: Institut Municipal del Paisatge Urbà i la Qualitat de Vida.

Altaió, V. (2000) *Desglossari d'un avantguardista*, Barcelona: Destino.

Benjamin, W. (1969) "The Work of Art in the Age of Mechanical Reproduction" in *Illuminations*, New York: Schocken Books, pp. 217–51.

Benjamin, W. (1973) *Charles Baudelaire: A Lyric Poet in the Era of High Capitalism*, London: NLB.

Bohigas, O. (1985) *Reconstrucció de Barcelona*, Barcelona: Edicions 62.

Bohigas, O. (1989) *Combat d'incerteses. Dietari de records*, Barcelona: Edicions 62.

Borja-Villel, M. J., Chevrier, J.-F. and Horsfield, C. (1996) *La ciutat de la gent*, Barcelona: Fundació Tàpies.

Boyer, C. M. (1994) *The City of Collective Memory, Its Historical Imagery and Architectural Entertainments*, Cambridge, MA: MIT Press.

Cia, B. (1999) "Acuerdo municipal para el proyecto del Besòs, en el que se invertirán 170.000 millones" *El País*, Sección Cataluña, 4 December, 7.

Cohen, P. (1998) "In Visible Cities: Urban Regeneration and the Local Subject in the Era of Multicultural Capitalism" in Beverley, J., Cohen, C. and Harvey, D. (eds) *Subculture and Homogeneization/Subcultura i homogeneització*, Barcelona: Fundació Tàpies, pp. 93–123.

de Certeau, M. (1985) "Practices of Space" in Blonsky, M. (ed.), *On Signs*, Baltimore: The Johns Hopkins University Press, pp. 122–45.

Duncan, C. (1991) "Art Museums and the Ritual of Citizenship" in Karp, I and Lavine, S. D. (eds), *Exhibiting Cultures. The Poetics and Politics of Museum Display*, Washington and London: Smithsonian Institution Press, pp. 89–103.

Espada, A. (1995) "Una pàtria: la ciutat dels JJOO" in García Espuche, A. and Navas T. (eds), *Retrat de Barcelona*, Vol. II, Barcelona: Ajuntament de Barcelona, CCCB, Institut Municipal d'Història, pp. 132–40.

Esquirol, J. M. (1998) *La frivolidad política del final de la historia*, Madrid: Caparrós Editores.

Etxezarreta, M., Recio, A. and Viladomiu, L. (1996) "Barcelona: Una ciudad extravertida", in Borja-Villel, M. J., Chevrier, J.-F. and Horsfield, C. *La ciutat de la gent*, Barcelona: Fundació Tàpies, pp. 280–89.

Fancelli, A. (1999) "Tony Blair adopta el 'modelo Barcelona'" *El País*, La Cultura, 4 July, 12.

Frisach, M. (2000) "Manuel Borja-Villel: 'El Macba ha de ser l'anti-Guggenheim'" *Avui*, 3 July, 34–5.

García Espuche, A. (1995) "Imatges i Imatge de Barcelona: una síntesi i tres epílegs" in García Espuche, A. and Navas, T. (eds), *Retrat de Barcelona*, Vol. I, Barcelona: Ajuntament de Barcelona, CCCB, Institut Municipal d'Història, pp. 41–62.

Gomà, R. (1997) "Degradació, crisi urbana i política de regeneració a Ciutat Vella de Barcelona, 1976–1993" in Institut Municipal d'Història de Barcelona, *El municipi de Barcelona i els combats pel govern de la ciutat*, Barcelona: Proa, pp. 285–307.

Grup d'Estudis Territorials i Urbans, del Departament de Geografia Urbana de la Universitat de Barcelona: Carreras i Verdaguer, C, Martínez i Rigol, S., and Romero Gil, J. (1999) *Els eixos comercials metropolitans*, Barcelona: Ajuntament de Barcelona.

Harris, N. (1990) *Cultural Excursions: Marketing Appetites and Cultural Tastes in Modern America*, Chicago: University of Chicago Press.

Harvey, D. (1989) "Flexible Accumulation through Urbanization: Reflections on 'Post-Modernism' in the American City" in *The Urban Experience*, Oxford: Blackwell.

Harvey, D. (1996) "The Social Construction of Space and Time" in *Justice, Nature and the Geography of Difference*, Oxford: Blackwell, pp. 210–47.

Irving, M. (1999) "Now that's what I call a city" in *The Guardian*, 26 June, 5–6.

Jameson, F. (1997) "Absent Totality" in Davidson, C. D. (ed.), *Anybody*, Cambridge: MIT/Anyone Corporation, pp. 124–31.

Jameson, F. (1991) "The Cultural Logic of Late Capitalism" in *Postmodernism or, the Cultural Logic of Late Capitalism*, Durham: Duke University Press, pp. 1–54.

MacCannell, D. (1976) *The Tourist. A New Theory of the Leisure Class*, New York: Schocken Books.

MadeinBarcelona (1999) (unpublished brochure).

Maragall, P. (1986) *Refent Barcelona*, Barcelona: Planeta.

Maragall, P. (1991) *La ciutat retrobada*, Barcelona: Edicions 62.

Moix, L. (1994a) *La ciudad de los arquitectos*, Barcelona: Anagrama.

Moix, L. (1994b) "Entrevista a Josep Ramoneda, director del CCCB: Es hora de hacer una revisión crítica de la modernidad" in *La Vanguardia*, 20 February, 48.

Ramírez, J. A. (1992) "La persistencia de los '*grands chantiers*' y los nuevos edificios rituales. Una crónica de París" in *Arte y arquitectura en la época del capitalismo triunfante*, Madrid: Visor, pp. 173–82.

Ramoneda, J. (1989) *La ciutat de les ciutats. Proposta d'equipament cultural per la Casa de Caritat de Barcelona*, unpublished dossier.

Ribalta, J. (ed.) (1998) *Servicio Público. Conversaciones sobre financiación pública y arte contemporáneo*, Salamanca: Ediciones Universidad de Salamanca.

Ricart, M. (2000) "El 2004 toma cuerpo. El Fòrum esboza un programa para atraer a 10 o 15 millones de visitantes" in *La Vanguardia*, Vivir en Barcelona, 25 February, 1.

Ricart, M. and Aroca J. V. (2000) "Besòs multimillonario. La operación urbanística del 2004 requerirá una inversión de 170,000" in *La Vanguardia*, Vivir en Barcelona, 4 December, 1.

Rifkin, J. (2000) *La era del acceso: La revolución de la nueva economía*, Barcelona: Paidós.

Ripalda, J. M. (1999) *Políticas postmodernas. Crónicas desde la zona oscura*, Madrid: Los Libros de la Catarata.

Roca, J. (1994) "Recomposició capitalista i periferització social" in *El Futur de les perifèries urbanes*, Barcelona: Institut de Batxillerat Barri Besòs-Generalitat de Catalunya, pp. 509–788.

Smith, N. (1996) *The New Urban Frontier. Gentrification and the Revanchist City*, London: Routledge.

Subirós, P. (1989) "Una ciutat per a la creació" in *El País*, Quadern, 30 March, 6.

Terdiman, R. (1993) *Present Past. Modernity and the Memory Crisis*, Ithaca and London: Cornell University Press.

Vázquez Montalbán, M. (1987) *Barcelonas*, Barcelona: Empúries.

Walsh, K.(1995) *The Representation of the Past. Museums and heritage in the post-modern world*, London: Routledge.

Wintour, P. and Thorpe, V. (1999) "Catalan cool will rule Britannia" in *The Observer*, 2 May, http://www.guardian.co.uk/Archive/Article/0,4273,3860699,00.html

http://www.barcelona2004.org/

http://store.yahoo.com/award-schemes/ribroygolmed.html

Notes

1. http://store.yahoo.com/award-schemes/ribroygolmed.html
2. See also Irving (1999: 5–6) and, in Spanish, Fancelli (1999: 12).
3. In calling urban spaces "texts, I do not mean that urban spaces constitute a discourse without extra-textual materiality. I mean that circulation in the city is possible only in as much as urban spaces are, semiotically speaking, made up of readable signs whose meanings are continuously negotiated and decoded. Phil Cohen (1998: 95–100) rejects this use of the concept, arguing that the metaphor of the city as text has historically allowed power structures to control the decoding process, stigmatizing as other anybody considered to be not properly "readable". Without denying the value of this critique, I believe that a semiotics of the city as text can still be politically useful. And in calling urban spaces "vehicles of ideology", I do not mean to imply a one-directional or static process. Different social actors, operating under conditions of inequality, can intervene to modify this text or to subvert its dominant uses (de Certeau 1985: 122–45).
4. Spaces are always socially constructed and subject to change and transformation. As Lefebvre put it, architecture is not an object but a process. Buildings, streets, plazas are

conditioned in their form and meaning by those who designed and financed them and, in so doing, endowed them with dominant meanings. But at the same time, they are lived, experienced and transformed in a myriad of spatial practices by the subjects interacting with them while using and circulating through these spaces. Once again, my argument does not deny the importance of these processes of signification, although it does not focus on them.

5 In the words of Manuel Borja (1996: 120): "[Aquestes citacions històriques] ens mostren la reconstrucció de la imatge, però no el drama [...] No sols no reconeixen el teixit humà i urbà de la zona, sinó que l'amaguen." Also, against this politics of amnesia, see Roca (1994: 730–32).

6 The concept of an architectural quotation of the past that is void of any meaning, also known as "historicism", is one of the central features of postmodern architecture. See Jameson (1991: 18). In a more general sense, the instability and infinite slippage of the signifier with respect to the signified is a constitutive tenet of post-structuralism. My criticism is not of the instability of the sign, but of the uses made of this instability to create a dominant "semantic politics" in democratic Barcelona.

7 Public space is nominally for all citizens, but in practicality access to it is restricted to certain individuals and groups. During the modern period, access to public spaces was determined by gender (since women — that is, bourgeois women — were supposed mainly to occupy the private sphere), but also by race and social class. According to Fredric Jameson (1997: 129): "private (space) is simply what interpellates me as an intruder while the existence or the waning of a truly public space can be measured by the degree to which it still interpellates me as a citizen". The key political question here is whether all individuals are really interpellated by public spaces and whether they have equal free access to them. For a history of the concept of public space, see Boyer (1994: 7–11).

8 Sculptures by Bryan Hunt, Oldemburg, Pau Gargallo, Tàpies, Chillida or Miró are still today dotted across peripheral or popular areas such as the north end of the Túnel de la Rovira, Parc del Clot, Can Dragó, Parc de la Creueta del Coll, Horta or Sants. Also characteristic of the architecture of this first Socialist period are the so-called "plazas duras" — squares built on a cement ground and very inexpensive to maintain and furnish — and the "casales" (civic centres) built in every city district.

9 See the City Council manifesto *Pla Estratègic Barcelona 2000*.

10 See Ferran Mascarell's words in Ribalta (1998:108). Mascarell is City Council Head of Culture since 1999 and before that was Head of the Institut de Cultura de Barcelona. He is an important figure in the formulation of the Socialists' cultural policies. See also Gomà (1997) and Etxezarreta *et al.* (1996).

11 Oriol Bohigas (1989: 125–26) explains in his memoirs that Narcís Serra, first social-democratic Mayor of post-Franco Barcelona, had since the early 1980s hoped that the city would be chosen for the Games because he saw this as a unique opportunity to secure its fast-track restructuring. This was not the first time that a big cultural-entertainment event had been utilized in Barcelona as an excuse for urban transformation in times of crisis and economic recession: the other two examples are the World Fairs organized by the city in 1888 and 1929.

12 The coincidence helps explain the process of co-option/institutionalization of the once very belligerent residents movements that would take place in the course of the 1980s (Roca 1994: 601).

13 Though he himself, in the successive institutional posts he held under Mayors Narcís Serra and Pasqual Maragall, was very instrumental in producing the opposite of what he had once preached (Bohigas 1989: 125–35).

14 A number of grassroots groups whose members are affected by these specific problems have been organizing to defend citizens' rights and to minimize the impact that these transformations are having on the city's social fabric and politics. Good examples of this are the Plataforma Cívica opposing the Projecte Barça 2000, and the Fòrum Ribera del Besòs.

15 In which he included Zaragoza, Toulouse, Montpellier, Valencia and Mallorca.
16 I am using the word "nationalist" here in its most common meaning in the Spanish political context, where it is applied to those ideologies whose political priority is the defence of the rights, however these are understood, of the historical nationalities within the Spanish state. Needless to say, parties covering the whole spectrum of the Spanish state, and specifically the PSOE, are not devoid of nationalism, Spanish nationalism in this case. My point is that Maragall's project for Barcelona, even though he ultimately belongs to the PSOE, is not fundamentally aligned to a Spanish, and clearly not to a Catalan, nationalism.
17 Which does not prevent their interests and objectives from often coinciding. On the one hand, the nation-state framework has proved very compatible with this politics of facilitating the leading role of cities. What we could call the Spanish state "politics of 92", fundamentally promoted three urban enclaves, Seville–Madrid–Barcelona. 34% of the one billion pesetas of public investment in the Olympic Games came from the state. On the other hand, the state-wide political and economic crisis following the carefully marketed boom of 1992 caused a decline in the city's prominence, because it signified a de facto halt to subsidies and investments in the most crucial and ambitious projects for the development of post-industrial Barcelona: the expansion of El Prat airport, and the building of a high-speed railroad line connecting Barcelona with Madrid and France (these projects are only now, in 2001, being materialized). Finally, the Catalan autonomous government has also promoted the internationalization of Catalan cultural and economic connections within the framework of the European Community, invoking a historic national community with south-eastern France: Catalunya Nord and Occitania. Joan Roca (1994: 607–8) notes the difference between the local (Barcelona) and autonomous government's strategies in this respect, arguing that the City Council proposes a community among cities, while the Generalitat does it among regions.
18 It could be argued that the urbanism and civility invoked by the Socialists in the 1980s has an important precedent in early twentieth-century Noucentiste discourse, particularly in terms of its anti-nationalist (Catalanist) implications, which both share. In this chapter, I limit myself to highlighting the ways in which the civility invoked in the 1980s was a response to contemporary historical conjunctions.
19 (Barcelona 2004, *10 preguntes sobre el Fòrum*. No date provided). This text could also be found on an early version of the Fòrum´s official web page that does not exist any more. After the advent of the anti-globalization movement phenomenon, and in particular after its local manifestation in Barcelona in June of 2001, a change in the rhetorical description of the Fòrum's aims and principles can be appreciated. These are now explicitly based on the Universal Declaration of Human Rights and the working principles of the United Nations (UNESCO is one of the Fòrum's sponsors). From the time of its first formulations by Pasqual Maragall in 1997, the project received numerous criticisms for its lack of specificity and focus. See Ricart (2000).
20 An acute criticism of the Fòrum project and the economic interests involved in the redevelopment of the areas where the Fòrum is supposed to be located (including parts of a marginal working-class neighbourhood, Besòs-La Mina) can be found in (MadeinBarcelona, 1999). See also Ricart and Aroca (2000).

Chapter 13

Changing course? Principles and tools for local sustainability

Enric Tello

Discovering the Mediterranean

"The creative, multifunctional city, which is also the most habitable city, is the one which contaminates least." This and other passages from the *Green Book on the Urban Environment*, presented to the European Commission in July 1990 (COM (90) 218 final: 30), might be considered as marking the birth of the new socio-ecological urbanism of the end of the twentieth century, which broke with the functionalism that had held sway since the years of the Letter of Athens (1933) inspired by Le Corbusier. Ironically, at a time when European urban planners (Krier 1993; Martínez Alier 1999) and even their American counterparts (O'Meara, 1999) are beginning to rediscover the environmental and social advantages of the traditional urban system of the Mediterranean, characterised by its dense, mixed, multifunctional design, our cities are undergoing an unprecedented metropolitan explosion, which is taking us further and further away from this ideal model (Nello 1998a,b; García Espuche 1998; Harvey 1998).

If we take the urban area of Barcelona as an example, in the past few decades almost as much land has been urbanized as in the preceding two millennia. In 1957, some 10,000 hectares had been urbanized; in 1972 this had grown to 20,000; in 1986 to some 40,000, in 1992 around 46,700 and two years later the area of urbanized land stood at about 50,000 hectares (Acebillo and Folch 2000). The rate of growth might have slowed down a little, but still *each year* an area greater than the whole of Barcelona's nineteenth-century plan of expansion (L'Eixample), a project designed in 1859 by Ildefons Cerdà, is urbanized (Rueda 1995; Rueda *et al*. 1998). Unlike other periods in history, this urban growth is directly *residential* in nature. It is totally unconnected with the location of new economic activities that attract residents. Neither does it have anything to do with demographic growth. In one of the regions with the lowest birth rates in the world, the population has remained stagnant at 4.25 million for over 20 years.

What lies behind this building fever? Between 1981 and 1996 the Barcelona municipality lost 240,000 residents from the 1,750,000 that it had had. If we include Badalona, l'Hospitalet, Cornellà, Santa Coloma and the other neighbouring towns in the *comarca* (administrative division including a number of municipalities), the loss reaches the figure of 325,000 residents. The vast majority have headed towards the new constructions in the second metropolitan ring, or have converted second homes into their main residence. In absolute terms the same people today occupy much more space, both within their homes and outside them. The result has been to increase the distance between the workplace and the place of residence, to multiply daily commuting statistics and to increase the segregation of people and activities in the territorial space.

Land prices and housing speculation have channelled these migratory flows in different directions according to income levels. The rich have moved into the more exclusive zones as they seek larger and more luxurious housing for the same price as in the city centre. The poor have been forced to move into municipalities where housing prices have remained more affordable. But neither their jobs, nor shops, nor public or private services have moved together with them. To meet their daily needs they must travel greater distances from places where the urban density and the provision of public transport are much lower. Commuting and dependence on the car have soared. The ratio between jobs and the population employed in the same municipality has gone into free fall. Quite the contrary to what the *Green Book on the Urban Environment* recommends. Why has this come about?

Voting with their feet

One thing is clear from the outset: if the total population has not increased, the residential demand comes to depend instead on the number of people living in the same house. Despite the general postponement of the age of emancipation among young people, as a reaction to the rising prices of housing and job insecurity, the number of residents per house has fallen throughout the metropolitan region from 3.9 in 1970 to an average figure of 2.9 in 1996. In the municipality of Barcelona there are 2.6 persons per house, and in 45% of the houses there are only one or two occupants. This reduction in the number of residents per house is, in turn, the consequence of the variation in the proportions between the different age groups. The change in the relative number of young and elderly people lies behind this apparent paradox that sees a stagnating population consuming more and more land to build new homes.

The emancipated youth constitute the principal source of demand for housing. Among the most elderly groups at the top end of the pyramid, the survival rates are beginning to fall with the passing of the years, and deaths see flats and houses being left vacant and coming on to the housing market. The net housing demand will be greater than that of the supply, or vice versa, according to the relative size of these two groups, the emancipated youth and the elderly (Wallace 2000). In the centre of cities such as Barcelona, the population pyramid is markedly aged. The age groups between 30 and 70 form a rectangular shape, occupying virtually all available housing. The highest death rates are recorded between 75 and 90 years of age, age groups that are still quite low in number. The cohorts born in the last *baby boom* of the twentieth century, between 1960 and 1980, generate most of the new demand. The former relinquish very little housing in relation to what is needed by the second group so that they can become emancipated from their parents.

This, among other reasons, explains why so many people are moving out to smaller municipalities in the second metropolitan ring where there is a greater supply of housing at more affordable prices. The population pyramids here are younger, reflecting this injection of new residents at the age of emancipation (Nello *et al.* 1998). A change in the family life-cycle is the main reason given when people are asked why they have moved. The myth of the house (as opposed to a flat in a shared block) has a stranglehold over the collective imagination: 70% considered this to be their ideal home in the latest metropolitan survey conducted in 1995 into the life and habits of the population. But it is not true that the majority of people move home in order to see their American dream of living in a detached house come true. When asked in which *comarca* the quality of life is best, *the majority still cite the name of the one in which they currently reside.* The percentage of people satisfied with their neighbourhood or city has increased in recent years.

Our culture remains a Mediterranean culture, with a tendency for its people to establish strong roots. If, despite this, so many people have had to uproot themselves, it has been because the inertia of these demographic factors has combined with urban speculation, and with growing inequalities in income and job opportunities, producing a powerful mechanism that ejects people from the city centre to ever more distant points in the metropolitan hinterland. In Barcelona, the average housing prices in 1991 represented approximately *fifteen times* the mean gross annual income per capita, and about *seventeen times* that in 1993. In some districts today the purchasing price is now more than *twenty times* the mean annual income per capita.

This price rise has nothing to do with a shortage of land on the market. By 1992, 46,700 hectares of the metropolitan region were urbanized, and there was a reserve of land designated for urbanization of 33,000 hectares (Acebillo and Folch 2000). Rather it has been caused by an added demand for this land: namely, the demand generated by those people seeking to invest their savings in property, the value of which seems certain to rise. Those hunting for a house in which to live therefore must compete with that proportion of savings that are invested in property, a proportion that is currently being generated by that third of all households which at year end do not spend all that they earn.

When urban land is subject to a purchase agreement, the house buyer has to buy two things: the house's usable space and the net worth of the land. The greater the flow of savings seeking to buy into the housing market, the more the demand for urban land grows above and beyond that generated by the mere need for housing. However, the final supply of flats and houses on the market will tend to adjust to the real demand for new housing. This makes land prices more expensive, and has repercussions on the price of the actual house, as long as there are always buyers prepared to pay the increased price of the building plot included within the housing price (and as long as the power on the market of both parties, expressed by their capacity to "wait", remains so uneven). Few investments are considered as safe as the purchase of property. This residential growth has increased differential incomes, while the inflexibility in the real demand for a basic staple good keeps investors' confidence high in the investment value of land.

Like an earth tremor, this wave of residential growth has spread in the 1980s and 1990s from its epicentre in the middle of the city causing the upheaval of the successive metropolitan rings. However, everything seems to indicate that at the start of the twenty-first century this tremor is about to go into remission. Over the coming decade, as that generation which was greatly reduced by a falling birth rate reaches the age of emancipation, the situation seems set to change. Today, for every person between the ages of 20 and 29 resident in some point of the metropolitan region there is another over the age of 65. In 2010 there will be *two*. In Barcelona itself the population is even older and the change in tendency will start to make itself felt earlier, leading to a "clearing out" of houses that could lead to a new migratory influx from the second ring to the city centre. Housing prices between the rings are already showing a tendency to converge, reflecting more the similarity in class of the residents of these areas than their particular location in the centre or in the periphery. This is exactly the same as what has happened in the cities of North America and the UK, where the effects of the new phase of demographic "implosion" are already making themselves felt (Wallace 2000: 82–4; Harvey 1998:113–14).

The drift towards unsustainability

In the meantime, the building fever of the past few decades of the twentieth century has bequeathed us an enormous set of social and environmental problems. In partnership with a model of mobility based on the automobile, it has sparked off a process of diffuse, socially segregating metropolitanization which is eroding that very diversity and mix of juxtaposed land uses that is so typical of our Mediterranean cities. It has eaten up huge quantities of fertile land, often of the very best quality, to the detriment of the non-urbanized open spaces whose environment has then been left to deteriorate. But its environmental impact is not only limited to the *footprint* it has left on the surrounding land. Within the detached houses and the scattered urban developments it also propitiates a rapid consumption of all types of natural resources — water, energy, a range of materials — and the generation of increasingly unsustainable waste that multiplies its global "ecological footprint".

It might be the case that the worst of its legacies has much in common with the perceived interrelation between the natural and the social worlds. The most harmful socio-environmental effects lie in the spread of what we might call the "commuter syndrome".[1] This mobility based on the car tends to dissolve the effective ties with the natural and social environment in which we live, and to corrode the democratic sense of community in which we commit ourselves to other people in the present and into the future. Our living space is coming to be identified less and less with the place in which we live, or where we return to sleep at night.

This syndrome leads to a growing polarization in the community between those who have a greater capacity to choose where they wish to live and work, and who are more able to travel by car or aeroplane, on the one hand; and all those persons, of whatever social class, gender, age group who either walk or use collective transport because the distance between the place in which they conduct their daily activity and place of residence is still close. The latest survey conducted in 1995 into the life and habits of the population of the metropolitan area of Barcelona reveals that "in general, men, young people and social groups with the highest level of education and professional standing have a greater capacity to move around the territory than the rest of the population: they travel further, they travel more frequently in private vehicles and proportionally invest less time" (Nello *et al*. 1998: 31).

Taken to their extreme, these tendencies imply that people will end up sharing their living space with their peers. This would mean the destruction of the city as a place where people can come together and communicate, with its integrative capacity for human development, for economic redistribution, conflict resolution and the construction of the basic sense of *citizenship*. The loss of diversity and mix is fatally linked to the withering of any sense of identification with a place or its people. The commuter syndrome is particularly disastrous when it comes to putting a stop to the environmental degradation that this very lifestyle is directly responsible for.

A conclusion that was more or less unanimously upheld by a workshop of experts gathered at the Centre of Contemporary Culture of Barcelona in 1998, during the exhibition entitled *The Sustainable City*, was the close relationship between the erosion of social cohesion and the environmental deterioration that was apparent as a result of the segregation of people in the metropolitan area induced by dispersed conurbation (Nello

1998b; see also *El País*, Catalan edition, 6/7/1998). It is not a purely local phenomenon. It is being reproduced everywhere, fostered by the impact of economic globalization, by the increase in inequalities of all kinds, and by the unsustainable rise in mobility based on the car, the aeroplane and the high-speed train (Fernández Durán 1993). Several reports undertaken for the European Commission point to similar conclusions. If social and political actions fail to change the trend towards segregation which sees the world being divided into winners and losers, the result in the long run will be the destruction of the city as a place where democratic citizenship takes root and makes itself manifest. The legitimization of the very process by which Europe can be built is at stake (European Commission 1997; Ministry of Development 2000).

Sooner or later many of these trends will change, even if nothing is done to change them. But if governments and the citizens themselves fail to intervene in time, the process that they feed will have had a deep, long-lasting social and environmental impact on the model of city. This is the context in which the movement of cities and towns for sustainability acquires its true meaning.

From the local to the global scale

In its origin, this movement in favour of local sustainability does not stem from the diagnosis of a drift towards unsustainability within each city and community brought on by capitalist globalization. The starting point is another: the new culture of *global* sustainability. This is the emerging awareness that we all live on "a single planet" and share a common destiny, a viewpoint that came of age at the Rio de Janeiro "Earth Summit" in 1992. This end-of-the-century globalization is a process operating on two fronts — the offspring of a schizoid culture. On the one hand, there is the dominant discourse of the past two decades concerning the inevitability of an unstoppable process of capitalist globalization, without any possible alternative response and, on the other hand, the general acceptance of an environmental crisis of planetary proportions that is threatening the future of mankind, and which demands that the development model that is causing the ecological deterioration be overhauled (Naredo and Valero 1999).

The result is patent: increasingly we speak of an environment that is in a worse and worse condition. This escalating rhetoric is a sign of our times. But the signs are not going totally unheeded, since the emerging culture of sustainability is already changing the framework of social legitimation. The existence of the environmental problem is not questioned. The weight of the evidence has shifted slowly onto the governments, business and the technocrats, who for the first time have to argue for the "sustainability", real or apparent, of their acts and products. The hegemony of the earlier productive mentality, which was so sure of itself in the "golden years" from 1950 to 1970, has been broken.

But among the clouds of rhetoric, the process of resolving matters stagnates (King and Schneider 1991). We know now that the situation is unsustainable. We are beginning to understand better how we should go about rectifying the situation. The bridges that have to be crossed to attain greater sustainability are today more realistic and specific than 25 years ago, when the ecology movements in the North and South were taking their first unsteady steps (Tello 2000). However, the obstacles that are being erected to these changes are formidable. We are living under siege, immersed in a situation in which the seeds of an ever-deepening crisis of legitimation are beginning to flower.

In the middle of this global political siege, the local action being taken by the movement of cities and towns for sustainability is giving cause for hope. It is worth stopping and asking ourselves why this should be so. I believe there to be two responses: the importance of *proximity*, and the clarification of the *mediative* actions that connect our small local world with those enormous, distant global processes that induce in the majority a great sense of powerlessness. Both remit to a common factor: it is not enough just to speak of the necessary socio-environmental changes. To be effective they need to be embodied within a community that takes responsibility for its own future. In addition to invoking a project, the agent that will make it a reality has to be constructed.

Proximity: participative democracy

Although not without its precedents, the movement of cities and towns for sustainability was created in Chapter 28 of the agenda for the twenty-first century, signed at the "Earth Summit" in Rio de Janeiro in 1992 (Keating 1996):

> Because so many of the problems and solutions being addressed by Agenda 21 have their roots in local activities, the participation and cooperation of local authorities will be a determining factor in fulfilling the objectives of sustainable development. Local authorities construct, operate and maintain economic, social and environmental infrastructure, oversee planning processes, establish local environmental policies and regulations, and assist in implementing national and regional environmental policies (Agenda 21, Chapter 28.1, "The local authorities", 1992).

But its main feature is described further on:

> As the level of governance closest to the people, they play a vital role in educating, mobilizing and responding to the public to promote sustainable development [...] By 1996, most local authorities in each country should have undertaken a consultative

process with their populations and achieved a consensus on 'a local Agenda 21' for the community (Agenda 21, Chapter 28.1, "Basis for action"; and 28.2, "Objectives", 1992).

Time has shown that this *proximity* makes a crucial difference. At a meeting held at the beginning of June 2000 in Malmo, Sweden, a hundred environmental ministers delivered a singularly unexpected declaration to the rest of the world: they recognized that they were failing in their efforts to stop environmental degradation. "There is an alarming discrepancy between the commitments made and the action undertaken", they said. "The main threats against nature identified at the summit in Rio de Janeiro in 1992 have become globalized and have been made worse because of the unsustainable consumer habits of the West" (*El País*, 1/6/2000).

This was not the first time that voices had been raised in opposition to the rhetoric of sustainable development (revered in all parts for its political correctness), accusing it of failing to provide concrete responses. What was new in Malmo was that this was readily admitted by the very ministers with responsibility for coming up with a remedy. To some this was clear incompetence. But this admission was exactly what was needed to begin to move sustainable development out of the limbo of well-meaning rhetoric into the harsh, conflictive reality of everyday life. It is not within the power of a single ministry, not even within that of all the governments of the world put together, to steer those trends that are pushing with such force in the direction of environmental deterioration away towards sustainability. Neither is it a task for those masquerading as Titans or dishonest tribunes. Quite simply, it is our common task. Translated into action, the new culture of sustainability is an invitation to participate.

The cities are in the best position to break the socio-environmental siege because they are the only space sufficiently local, and at the same time global, in which a more participative democracy is being witnessed. While the national Agenda 21s have come to nothing, and the international agreements that arose out of the global Agenda 21 signed in Rio ran into one obstacle after another, the movement of the cities and towns for sustainability has set out on a path that is undoubtedly full of hurdles and contradictions, but one which is also highly revealing. Indeed, this movement cannot be dismissed as trivial if we recall the words from the preamble to the Aalborg Charter that *80% of the European population lives in urban areas*. At the global scale, too, most of the population now lives in urbanized areas or metropolises. The twentieth century has been, among other things, the century of the spread of urbanization to incorporate the majority of mankind. Either we open up practicable paths towards sustainability in and from the cities, or the global socio-ecological problem is without solution (Harvey 1998).

The "Charter of the European Cities towards sustainability", signed in the small Danish town of Aalborg in 1994, adopts a radically different position from that of the state governments at international conferences on the environment. While intergovernmental negotiations founder, as one multilateral block follows another — "you make the first move" could be its motto — the movement of cities and towns for sustainability proposes taking the initiative unilaterally — *put your own house in order first* and then demand reciprocal actions from the others.

Coming back down to earth again

All the reports on the world situation show that environmental degradation and social polarization continue unabated (Brown *et al.* 2000 and 2001; Flavin *et al.* 2001; PNUD 2000; PNUMA 2000; World Resources Institute 2000). Local initiatives are not capable of changing the course the planet is taking. And it is still too soon to take stock of the actions and results of the incipient movement of cities and towns for sustainability. The acid test needs to be conducted among those cases where declarations are slowly being turned into decisions.

Notable among these, standing in sharp contrast to the stagnation surrounding the multilateral agreement to check climate change, is the unilateral initiative undertaken by the city of Heidelberg. In 1992, this small German university city decided to adopt the following *voluntary* commitment: to reduce its emissions of greenhouse gases to 80% of their 1987 levels by the year 2005. It is worth stressing that this 20% reduction was the first step that the two main coalitions of ecological groups of the North and citizen movements of the South — Climate Action Network and Third World Network — proposed to the governments of the OECD countries at the Kyoto summit in 1997.

The 20% proposal fell upon deaf ears. As is now well known, in Kyoto the developed countries agreed to an average reduction for the period 2008–2012 of an estimated 5.2% on the figures for 1990. The European Union agreed to a combined reduction of 8%, which allowed countries such as Spain to increase their emissions by 15% (by 1999 they had already increased their emissions by 27%, and soared until 35% in 2001; Nieto and Santamarta 2000; Santamarta 2000). Japan agreed to 6%, and the United States to 7% but then President Bush refused to ratify the Kyoto protocol. All the indications are that even those insignificant targets will not be met. This helps to put the importance of the initiative taken by Heidelberg in its true perspective.

Beginning by "putting one's own house in order" gives one, among other things, a legitimate right to demand a reciprocal response from the others. It also entitles one to adopt the mantle of leader. Having made public its 20% target, in 1994 Heidelberg convened a summit meeting of local authorities backed by the OECD and the EU with the aim of examining "how to combat global climate change using local actions". The objective of Heidelberg was adopted by the international campaign "Cities for Climate Protection", backed by the ICLEI, which had been set up following the 1992

"Earth Summit". Barcelona joined this campaign in 1993, and attended the 1994 Heidelberg Conference together with 150 cities from around the world with the aim of reducing the 1987 levels of greenhouse gas emissions by 20% by the year 2005 (Figure 13.1).

The decision taken by the city of Barcelona to join the campaign did not come about by chance. In 1993, the first public meeting on the urban environment had been held thanks to the popular initiative of a citizens' platform formed by ecology groups, neighbourhood associations and the leading unions under the name of "*Barcelona Saves Energy*". The number of members mobilized by these entities was greater than 100,000, which ensured that it did not go unnoticed in the media and that it had certain repercussions within local politics (Tello 1996). Since then, Barcelona City Council has been at the forefront of the world's cities in the signing of international agreements in support of global sustainability. For this reason, an analysis of the city's results should serve as an example of the current situation.

13.1
Photovoltaic panels on roof of Council House, late 1990s

Clarifying what needs to be done

For the time being, the signing of the Heidelberg commitment has done little more than reveal the quantity and the origin of greenhouse gas emissions in the city of Barcelona (ITEMA 1996). We have discovered three important things. First, when the "economy is going well", the environment is almost certainly suffering. Between 1987 and 1992, and again from 1996 onwards, the recovery witnessed in the economy has meant an increase in Barcelona's contribution to global warming (Table 13.1). The culture of eco-efficiency has, it is clear, yet to leave its infancy. We have still to learn how to increase human development while reducing our consumption of natural resources (von Weizsäcker *et al.* 1997).

Second, these data allow us to identify the origin of the emissions. If the city can be considered to be comprised basically of buildings and streets which interconnect with one another and with the exterior, the direct emissions of greenhouse gases can be divided into three main groups (see Figure 13.2). Those caused by the electricity and gas consumed within the buildings account for 37% of the total. The traffic which fills the streets generates 28%, a proportion originated in the main by private cars. The incorrect use of the waste that the city excretes every night is responsible for the other 36%, caused by the emissions of methane from the rubbish dump in Garraf (the destination, during many years, of two-thirds of the rubbish that has not been previously sorted) and the CO_2 from the incinerator in Sant Adrià (where the remaining third is burnt).

Third, these estimates allow us to evaluate just how far we are still from a situation that might be described as sustainable. We can

Table 13.1   Direct emissions of greenhouse gases in Barcelona (1987–96)

	Tonnes of CO_2 equivalent*	Number of inhabitants	Tonnes/inhabitant
1987	3,242,922	1,703,744	1.90
1988	3,331,298	1,714,355	1.94
1989	3,472,577	1,712,350	2.03
1990	3,562,727	1,707,286	2.09
1991	3,695,326	1,643,542	2.25
1992	3,948,959	1,630,635	2.42
1993	3,617,540	1,635,067	2.21
1994	3,569,696	1,630,867	2.19
1995	3,622,095	1,614,571	2.24
1996	3,824,524	1,508,805	2.53
1997	3,976,172	1,505,581	2.64
1998	4,072,151	1,503,451	2.71
1999	4,302,220	1,496,266	2.88

*   In addition to CO_2, these figures include emissions of methane from waste sites.

Source: based on data drawn from the *Pla de Millora Energética de Barcelona*, 2002

assimilate these 3 tonnes of CO_2 equivalent per inhabitant in 1999 within those originated by the city as such, and for which the city has a direct political responsibility. It should be remembered, however, that these figures only take into consideration the impact of the consumption of energy sources within the city, and the burning and dumping of waste originated by the city's metabolism. They do not include the emissions caused by the processing of the products that the inhabitants consume (food, cement, industrial products, water, etc.).

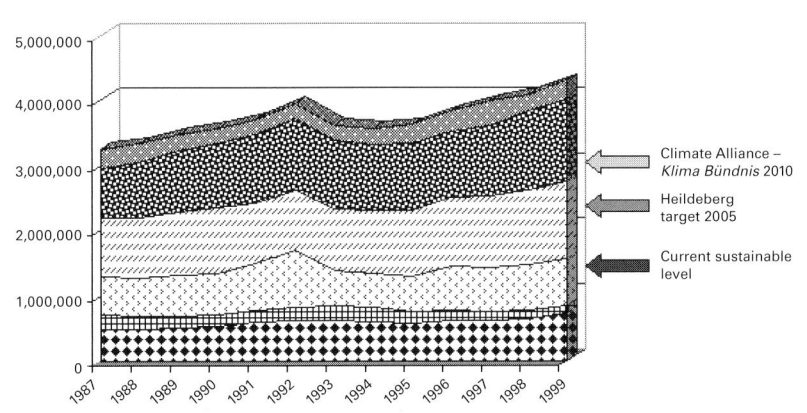

13.2
Greenhouse gas emissions in Barcelona 1987–99 (tonnes of CO_2 equivalent)

Getting the homework done

The first two arrows to the right of Figure 13.2 indicate the approximate values established as benchmarks by two agreements of which Barcelona City Council is a signatory. The first, subscribed to in Heidelberg in 1994, involves "reducing greenhouse gas emissions by at least 20% of 1987 levels by the year 2005". The second, established by the Climate Alliance — Klima Bündnis — of cities committed to climate protection, which Barcelona subsequently signed, proposes reducing these emissions by 27% of 1997 levels by the year 2010.

Although the two agreements are not mutually consistent, the benchmarks they set are not inconsiderable. The targets are ambitious but not impossible. The graph itself indicates the easiest way of achieving them: changing waste management policies by adopting two key actions. First, the rubbish dump site in Garraf should be shut down and the methane generated there recovered for use as a substitute for other combustible fuels (for example, gas oil for the city's fleet of buses, the natural gas of the network, and electricity). The project is up and running and an experimental plant is already generating electricity with a small part of this biogas. But its large-scale use depends on the progress made by the Metropolitan Programme for the Management of Municipal Waste passed in 1997, although this is already well behind schedule. The complete reform of the rubbish dump site, put off time after time, has now been proposed for after 2006.

The second step that would have to be taken to meet the Heidelberg objectives is the closure of the incinerator in Sant Adrià del Besòs. This has yet to appear on any political manifesto, and the celebration of the World Cultural Forum in 2004, with its overt message in favour of sustainability, in front of its very gates, will do little to lead to its closure. If these waste management measures are not implemented soon, only two outcomes are possible: either the city will fail to fulfil the commitments it has subscribed to, or it will have to reduce the emissions from the city's buildings and streets. This will mean major reductions in the use of the motor car, improvements to the energy efficiency of the city's buildings, and the development of large-scale solar energy projects. A recent up-date on a study commissioned to examine how Barcelona functions as an urban ecosystem shows that the amount of solar energy that goes unused each year in the city amounts to 47.5 GJ/ha, which is five times greater than all the primary energy obtained from fossil fuels and nuclear energy imported from abroad, and is nine times greater than its final energy consumption (Barracó *et al*. 1999).

If we consider the 276 million GJ/year of solar radiation that falls on the city's constructed area (not counting streets, squares and green zones) this is still 14 times greater than all the electricity consumed, and 46 times greater than domestic electricity consumption. Supposing a modest commercial yield of 10%, to produce all the domestic electricity that Barcelona currently consumes would require a photovoltaic coverage equivalent to 22% of its built area (1,259 ha). To supply all this, the surface required would be equal to 70% of the total (4,062 ha), although this area would not have to be so great were consumption to be more efficient. With the 51,000 kilowatt-hours generated each year by the 603 m² of thermo-photovoltaic cells, the Pompeu Fabra Library in Mataró shows that this is no technological pipe dream.[2]

The greatest disparity between the commitments made at the international conventions and the course taken by the realities of everyday are, beyond doubt, to be found in the third "packet", responsible for generating 30% of the CO_2 in Barcelona (and a much greater proportion of the local atmospheric contamination, which is also on the increase): the city's road traffic. Barcelona City Council is a founder member in Spain, together with Cáceres, Córdoba, Donosti, Granada, Jaén, Oviedo, Sabadell, Toledo and Málaga, of the EU sponsored network that carries the name of *Car Free Cities*.[3]

This club of local authorities that seeks to promote car free cities was founded in 1994, following an official declaration from the European Commission that considered this environmental aim a highly recommendable objective for economic reasons, too (COM(92)231 final, cited by Whitelegg 1993). Such a singular declaration, quite out of keeping with the dominant transport policies of the EU today (Fernández Durán 1997), arose out of a study commissioned in 1991 by the EU to calculate how much it would affect the euro in the average citizen's pocket, and the public budgets, to live in a car free city. In the words of Fabio Maria Ciuffini, director of the study (Ciuffini 1991: 182; Ciuffini 1993: 48–49):

> If the figures are calculated properly, that is, if the public and private costs of the urban transport system are summed, the city of the automobile can be up to five times more expensive than the city without cars, and that is before accounting for, or even beginning to express in terms of money (were this possible, although it is patently not) the environmental effects of the motorized city.

But the people of Barcelona are unaware of the fact that, on paper, theirs is a car free city. With the exception of the organization of a car free day each year since 1999, Barcelona City Council has not adopted any specific plan that seeks a significant reduction in the 600,000 internal journeys (20% of the total), and the 875,000 external journeys (52%) that are undertaken each day by car. More than 70% of the internal journeys, and 45% of the external ones, are undertaken on foot or on public transport.[4] But the roads, all of which are virtually destined for use by private motor vehicles, occupy two-thirds of public road network space. Those 70% who travel on foot or on public transport have to make do with the remaining third. Despite the minor changes made to rectify this situation since 1992, the space for pedestrians has only increased in small proportions. However, interesting proposals exist

13.3
The airport, with the expansion for 1992

such as the introduction of "super-blocks" of restricted traffic and priority for pedestrians (Rueda 1995: 140–42), which are gaining the active support of neighbourhood and ecological associations (Figures 13.3 and 13.4).

The distance to the horizon

Meeting the Heidelberg objectives by 2005 or those of the Climate Alliance — Klima Bündnis — by 2010, through the actual implementation of any of these proposals, would represent a major step towards sustainability, but we would still be some way from attaining sustainability itself. The city is not an island, and in addition to the direct emissions described above we have to remember those that are incorporated in the products made outside the city. The team responsible for drafting the Agenda 21 for Barcelona estimate that the latter are similar in magnitude if not greater than the direct emissions. This would mean that total emissions rise to some 6.5 tonnes per inhabitant each year.

This estimate is credible, given that in Spain this year the average emissions were around 7 gross tonnes of CO_2 equivalent per inhabitant, taking into account only CO_2 and methane. The dependence on nuclear power for electricity is not as great in Spain as a whole as it is in Catalonia, where lamentably it is over 70%, and the "backpack" of CO_2 attached to each kilowatt of electricity consumed is therefore greater. In the year 2000, the total greenhouse gas emissions, including N_2O, HFCs, PFCs and SF_6, had already reached 10 tonnes per person per year in Spain. If we subtract those absorbed in the local sewage channels, the net emission is around 9 tonnes (Nieto and Santamarta 2000). According to the Wuppertal Institute,

13.4
Coastal ring road, near Vila Olímpica, 1998

a sustainable level of emissions should not exceed the threshold of 2.3 tonnes of CO_2 per person per year (von Weizsäcker et al. 1997: 295). Those generated directly by the metabolism of cities such as Barcelona, whose economy is based around the service sectors, should tend towards 1 tonne per inhabitant per year.

Indeed, this is the third arrow drawn on Figure 13.2. It means that a sustainable Barcelona should cut its contribution to global warming by reducing its CO_2 emission levels to 1.5 million tonnes a year. This means cutting the rates for 1999 by 65%. These reductions are comparable to those recommended at the global scale by the Intergovernmental Panel on Climate Change (IPCC), and are within the orders of magnitude of the "Factor Four" strategy towards eco-efficiency: double global well-being with half the natural resources (von Weizsäcker et al. 1997). In the case of Barcelona, human development needs to dissociate itself considerably more from the consumption of resources than the global average. As far as CO_2 emissions are concerned, "factor four" would mean more or less the following for Barcelona: sustain the level of well-being with a quarter of the current levels of natural resource consumption.

Tracking the eco-footprints of unsustainability

Calculating the global "ecological footprint" results in similar orders of magnitude. Its size is inversely related to the efficiency of the city, and directly related to its scale of consumption. For each unit of end consumption, the greater the efficiency recorded, the lower this impact will be. According to studies currently being undertaken, the surface area of woodland needed to absorb the emissions of CO_2 usually represents between one-third and one-half of the overall ecological footprint of each country or city (Rees 1996; Wackernagel 1996; Wackernagel et al. 2000). In line with the standard formula, the "ecological footprint" is defined as the photosynthetically active surface needed to replace the high quality energy dissipated by industrial metabolism at a given point on the planet (Rees, 2000: 372). It tells us approximately how much biologically productive land and sea are needed to produce the resources that a given population with a certain technology and lifestyle consume, and to absorb its main waste products (Wackernagel and Silverstein 2000: 392).

This estimate of equivalent areas depends on a key factor: the carbon absorption rate due to the growth of the world's forests. This parameter varies from one ecosystem to another. The slow-growing Mediterranean sclerophyllous woods absorb 3.7 tonnes of CO_2 per hectare each year, compared to an average of 6.6 globally (Rees and Wackernagel 1996: 73; Fòrum Cívic Barcelona Sostenible, 1998). But having entered the atmosphere, the CO_2 molecules do not carry their corresponding "appellation d'origine". Calculations of ecological footprints and debts are based on the average global absorption rates because the function of recycling CO_2 is a global common good. This helps clarify our dependence as cities on what happens in the rest of the world. Home-based sustainability is no longer possible. It can only be achieved by signing a global pact.

An advantage of using the "ecological footprint" as an index of global sustainability is that, once calculated, we can compare it with the environmental space actually available on the planet. A deficit between the "average environmental space" and the ecological footprint per inhabitant reveals the existence of an "ecological debt", contracted by our city or country with those who record a positive balance. The size of the accumulated deficit then serves as another measure of a place's unsustainability. Breaking down the footprint into its various component parts helps to establish the magnitude of the task that awaits us and identify the priorities of this task: improving the efficiency and/or reducing the scale of consumption in various sectors and in varying proportions (Table 13.2).

To sustain the consumption of each person living in Barcelona requires 3.25 hectares per person, almost twice that of the 1.71 hectare benchmark available for each inhabitant on the planet. If we include 0.27 hectares per person to preserve the biodiversity (equivalent to 12% of the biologically active surface of the Earth shared out among the human population) Barcelona needs an overall area greater than 5 million hectares: 538 times the area of its municipality. 25% of the Catalan population "occupies" as much territory as 1.5 Catalonias. The size of this "ecological footprint" is similar to that estimated for Spain as a whole (although the latter is somewhat greater, because the most energy-intensive activities are usually located outside cities with economies that are predominantly operating in the tertiary sector such as Barcelona's, and also because Spain's dependence on nuclear electricity is, in general, not as great) (Table 13.3).

Table 13.2 **Barcelona's ecological footprint, and its global ecological debt (1996)**

	Barcelona's ecological footprint (ha/capita)	Total surface area required (ha)	Available space globally (ha/capita)	Barcelona's "ecological debt" (ha/capita)
Crops	0.49	739,314	0.25	−0.24
Woods	0.08	120,708	0.60	−0.50
Absorption of CO_2	1.02	1,538,981		
Pasture	0.99	1,493,717	0.60	−0.39
Sea	0.65	980,723	0.50	−0.15
Constructed land	0.005	7,544	0.03	+0.025
Protected land*	0.001[†]	407,377	−0.27*	−0.27
Total	3.23	5,288,364	1.71	−1.53

* Subtracting an overall 12% to preserve the biodiversity.
[†] Including only the 1,795 hectares of the Collserola natural park that forms part of the municipality.

Source: based on Ralea and Prat (1999: 25) and Wackernagel et al. (2000)

Table 13.3 **Ecological footprints in various countries around the world (1997)**

	Ecological footprint per inhabitant (hectares)		Ecological footprint per inhabitant (hectares)
Germany	5.3	Bangladesh	0.5
Australia	9.0	Brazil	3.1
Austria	4.1	Chile	2.5
Belgium	5.0	China	1.2
Canada	7.7	Colombia	2.0
Spain	3.8	Egypt	1.2
United States	10.3	Ethiopia	0.8
France	4.1	India	0.8
Great Britain	5.2	Mexico	2.6
Holland	5.3	Nigeria	1.5
Italy	4.2	Pakistan	0.8
Japan	4.3	Peru	1.6
Portugal	3.8	Thailand	2.8
Russia	6.0	Turkey	2.1
Sweden	5.9	Venezuela	3.8
World average	2.8		
Available biocapacity	1.7		

Source: Wackernagel *et al.* 2000

If the whole world lived like the people of Barcelona, or like the Spanish as a whole, we would need two planet Earths. If the whole world wished to live like the people of North America, with an ecological footprint greater than 10 hectares per inhabitant, we would need six. But we only have one. This gives us an accurate idea of just how unsustainable we are, and of the magnitude of the task that faces us. Reaching a situation that is more or less sustainable would mean, in the case of Barcelona, reducing its current footprint by half, adopting the "factor four" strategy. For the United States it would mean adopting a "factor ten" strategy, to reduce the size of the footprint to one-sixth of its present size.

As Mathis Wackernagel and his co-workers point out, the figures that make up the ecological footprint "reveal the extent to which the richest people and the richest countries have actually 'appropriated' the productive capacity of the biosphere". This has "unmistakable implications for equity" (Wackernagel *et al.* 2000). Not everyone accepts, however, that all the Earth's inhabitants have the same rights, both now and in the future, to use the capacity of carbon absorption of the natural common systems. When an official delegate from the USA heard a very moderate proposal for global convergence on the emissions of CO_2 per inhabitant during the first half of the twenty-first century, presented at the Kyoto Summit by the Global Commons Institute, he exclaimed: "this seems to me to be like Global Communism. I thought we had won the cold war …."[5]

City, not suburb

The crossroads are clearly sign-posted. So is the road towards eco-efficiency and environmental justice. But before we can set out along this road, and while we continue to crash into the obstacles that block our path, the metropolitan explosion of our cities is pushing us further and further towards unsustainability. If our cities today are ecologically unsustainable, then those suburban stretches that developed in the spread of the conurbation are even more so.

The growth of the residential wave witnessed over the past 20 years has multiplied the consumption of all types of resources (energy, water, land and materials) and has increased the generation of waste and the emission of contaminating gases both in the local environment and in the global atmosphere. Ironically, many of those who have moved to low population density urban zones, in places increasingly more distant from the metropolitan rings, hoped to find a better environment. But, like the plague in days gone by, they are contributing to its deterioration with a suburban model that is much more inefficient in its use of natural resources, and more socially unjust.

To illustrate this trend, Table 13.4 contrasts the figures for commuting, the consumption of energy and water, the emission of greenhouse gases, and the generation of waste in the municipality of Barcelona, with its 1.5 million inhabitants, and the nine municipalities of the sub-comarca of the Baix Maresme, which have seen their population grow from 52,734 in 1975 to 94,148 in 1996.

Table 13.4 **Comparison of the socio-ecological metabolism of Barcelona and a zone in the second metropolitan ring: Baix Maresme***

Parameters	Baix Maresme	Barcelona
Gross family income per capita (in pesetas — 1995 value — × 1,000)	1,954.3	1,799.4
No. of private cars per thousand inhabitants (1997)	472	415
Percentage of internal journeys-to-work/population employed in the municipality	50.8	78.8
Percentage of journeys on foot as proportion of daily internal mobility (1999)	44.3	36.0
Percentage of public transport as proportion of daily internal mobility (1999)[†]	8.4	38.0
Percentage of private vehicles as proportion of daily internal mobility (1999)[†]	47.3	26.0
Energy consumption/inhabitant in transport (TEP/capita in 1997)	0.62	0.36
Domestic consumption of electricity/capita (kWh in 1997–98)	1,641	1,098
Domestic consumption of natural gas/capita (therms in 1997)	1,725	1,169
CO_2/capita derived from final energy consumption (Tm, 1997–99)	3.84	1.98
Litres of water/inhabitant/day for urban uses (1996–97)	289	210
Litres of water/inhabitant/day for domestic uses (1996–97)	244	135
Kilogram/capita/day of municipal waste generated (1999)	1.6	1.4
Kilogram/capita/day of waste separated and taken to a recycling plant	0.098	0.150

[*] Alella, Cabrera de Mar, Cabrils, el Masnou, Premià de Dalt, Premià de Mar, Teià, Vilassar de Dalt and Vilassar de Mar.
[†] Includes taxis, which in Barcelona accounted for 4.6% of daily mobility of persons in 1994. Collective systems of public transport accounted for 36.5%. The remaining 1.6% includes bicycles and school and company buses. Of the 25.2% of internal journeys made in private vehicles, 19.9% were by car and 5.3% by motorbike or scooters.

Table 13.5 **Correlations between the consumption of natural resources, urban density and available family income in the municipalities of the Baix Maresme in the 1990s**

Correlation between ...	Gross available family income per inhabitant (pesetas in 1995) $R^2 =$	Urban density (inhabitants/ha of urban land, 1998) $R^2 =$
Number of cars per inhabitant (1997) and ...	0.1924*	0.7775*
Percentage of self-sufficiency in terms of commuting and ...	0.1173	0.4516
Consumption of domestic energy/capita (1997) and ...	0.1256[†]	0.5863[†]
Domestic electricity/capita (1997–98) and ...	0.1848[†]	0.7084[†]
Domestic consumption of water/capita (1999) and ...	0.0889[†]	0.6567[†]
Municipal waste/capita (1998)[‡]	0.2269[§]	0.5613[§]

* Not including the extreme case of Cabrera de Mar, where there were 0.968 private cars per capita (in this municipality the R^2 are: 0.0118 and 0.3243 respectively).
[†] Adjusted according to the percentage of second homes in the last housing census (1991).
[‡] Corrected in accordance with the seasonal population, estimated by the Waste Board.
[§] Not including the extreme case of Premià de Dalt, which only generated 0.74 kilograms/inhabitant a day (in this municipality the R^2 are: 0.0044 and 0.2108 respectively).

The amount of energy consumed per inhabitant in transport is 72% higher in the Baix Maresme than in Barcelona; while the energy used by homes and businesses is 15% higher. The rate of the emission of greenhouses gases per inhabitant linked to mobility is twice that recorded in Barcelona. The emissions derived from the domestic and commercial consumption of energy are 35% higher. The trends are clear, but their destiny is not necessarily a foregone conclusion. In the Maresme, the municipalities with the lowest urban density, the greatest proportion of scattered urban areas, and the most detached housing, also tend to have more cars per capita, to be less self-sufficient in terms of commuting, to consume greater amounts of energy and water, and to generate more waste per inhabitant than the denser and more compact towns. This shows that *the city model is much more important in determining the socio-ecological behaviour and its environmental impact than per capita available income* (Table 13.5).

These and other examples highlight the importance of an urban policy that needs to be implemented urgently in the municipalities of the

13.5
View of park next to river Besòs, showing powerlines (removed in 2003)

metropolitan rings that are suffering the effects of the latest residential wave. To stem these unsustainable trends in the urban peripheries a strategy needs to be adopted that encourages sustainability and strengthens the links of proximity, that fosters a community spirit and promotes a greater commitment with the people and the natural environment of the place in which it lives (Figure 13.5). If so many people are moving into these areas in order to be closer to nature, and to improve their quality of life, they must also understand that in order to enjoy it they have to look after it.

Taking root

It is here that the spread of the sustainable cities movement towards smaller towns and municipalities, and its joint actions, become most important. In Catalonia, the initiative has been taken by the "Network of Cities and Towns for Sustainability", with the backing of the Department for Environmental Affairs within the Diputació (Barcelona Provincial Council). By the end of 1999, the network was already made up by 140 municipalities, with a combined population of 4.5 million inhabitants (that is 74% of the Catalan population).[6] The "Manresa Declaration", signed in 1997, sets out the objectives of this network invoking the Aalborg Charter and the Lisbon Action Plan. The main objectives are to strive to link environmental issues with socio-economic and cultural matters,[7] to encourage citizen participation,[8] to interchange experiences, and to promote joint actions "that increase the relative weight of the cities, especially that of the medium-size and small cities, within the autonomous communities, the state and Europe, as a guarantee of local rebalance and sustainability".

Twenty years after the restoration of democracy in the town councils of Catalonia, the sustainable culture is beginning to take root and is bringing about a rebirth in local politics, as it has to face up to new challenges and to widen its horizons. It is a pity, however, that this interest should be resurfacing at a time when the building fever caused by the great residential changes of the past 20 years is on the point of subsiding. Yet, such a renewal is better late than never. After the storm, the task of constructing a polycentric system of cities that is more compact, diverse and socially integrated will be a mammoth one. This system will have to strengthen its own retaining walls, build an efficient public transport network that provides a real alternative to the private car, and capitalize on plans that manage the demand for water and energy so as to promote the creation of new jobs that are located locally and that are sustainable with a diverse, flexible and democratically controlled economic structure.

Before becoming a Network member, the municipality must ratify the Aalborg Charter and join the European movement of cities and towns for sustainability. To date, the Network's main task has been to encourage the Diputació to initiate the Agenda 21 process by undertaking municipal and supra-municipal environmental audits (Sureda 1999). As happened in the case of Barcelona's audit, such studies can provide highly useful information for making an accurate diagnosis, identifying priorities, and noting alternatives. As such they are a basic tool, both for the town councils and the citizens' associations.

Of the audits that have so far been completed, it is clear that the medium-size and small cities often suffer more severe problems, if this is possible, than the main central city. But equally they are sometimes in a better position to begin steps towards sustainability. For example, solar radiation per capita is 22 times greater in the Maresme (7,533 GJ/capita/year) than it is in the Barcelona Municipality (336 GJ/capita/year), which means in order to supply all the electricity based on solar power would only require a photovoltaic surface equivalent to 1.4% of the comarca (Acebillo and Folch 2000: 324). In the hydrological balance of the Baix Maresme, the precipitation (23 Hm^3/year — Hm^3 = 100 cubic metres) is twice that of the water brought in from outside the area (10.8 Hm^3/year). In Barcelona, local rainfall (60 Hm^3/year) accounts for just 40% of the supply from outside (153 Hm^3/year).

However, it happens to be the case that the levels of energy and water consumption per capita are much higher in those areas where the housing has the greatest capacity to exploit the rainfall and the solar radiation (roofs, size of the houses, purchasing power, etc.). In Barcelona the daily domestic consumption of water per person in 1997 was 135.3 litres (Barracó *et al*. 1999: 36–42). In the working-class towns of the city's industrial belt where incomes are lower, such as Badalona, Santa Coloma, l'Hospitalet and Cornellà, the average consumption is even lower: 114 litres per person per day (Esquerrà *et al*. 1999: 76–77). In municipalities with higher income levels than those of Barcelona, such as the Baix Maresme, domestic water consumption is beginning to run away with itself: ranging from 245 litres per person per day in Alella and Vilassar de Dalt to more than four hundred in Cabrera and Cabrils.

This is no trivial matter, if we realise that a considerable proportion of the estimated "hydrological deficit", which supposedly will require the massive transfer of water from the river Ebro or the Rhône to the metropolitan area of Barcelona, is based exactly on the extrapolation of these differences into the future. All it requires is an effective demand management programme to bring about a convergence of consumption levels in the second ring with those in Barcelona and its suburbs, for these alarming "hydrological deficits" to evaporate (Tello 1999: 33). To illustrate this, the flushing of toilets alone represents a minimum annual consumption per person of 5 m^3, equivalent to 20 Hm^3 throughout the metropolitan area.

An efficient strategy to manage the local demand for water and energy is not only a way of avoiding the transfer of growing environmental impacts to increasingly distant points. For the movement of cities and towns that wish to set out on the road to sustainability, saving water and harnessing the power of the sun represent a magnificent opportunity to overcome the "commuter syndrome" engendered by the massive shift in residences and the socio-ecological segregation of people in the territory.

If it's Tuesday, this must be Belgium

The dialectic existing between socio-environmental and economic globalization on the one hand, and the possibilities of strengthening the democratic control of the development model from within the local and regional spheres on the other, is essential in taking steps towards sustainability. As the environmental economist Richard Norgaard argues, economic globalization leads to an increase in the distance between those who take production and consumption decisions, and those who suffer their ecological and social effects. This growing distance multiplies the transactional costs when negotiating common solutions (Norgaard 1997: 184):

> The geographical expansion of exchange increases physical distance. At such distances, people are more unlikely to be able to perceive the consequences of their acts. Those that can see the results are to be found in one place, those that can do something about them are to be found in another, and the distance that separates them hinders communication and the reaching of a consensus on common solutions. The specialization that accompanies the increase in commercial exchange also increases social distance because it reduces shared experiences and common ways of seeing the world.

The solution to this "commuter syndrome" — this universal estrangement — must lie in a commitment to relocation: the establishment of a local orientation in economic, social and political networks. Instead of squandering a whole range of resources imported at increasingly greater costs (economic, ecological and social), public and private expenditure needs to be redirected towards the more efficient use of accessible environmental resources, such as the sun, the rain and fresh food produce, thereby fostering the local development of new sectors of economic activity. The design and engineering of new apparatuses and systems, the manufacture and installation of these goods for domestic, industrial and agricultural use, represent a huge potential for jobs that can be provided locally and which are sustainable.

The strategy of *proximity* should not be mistaken for adopting an excessively local or parochial perspective. After all, sustainability is by its very nature holistic. If each side to the problem contributes what it can to the solution, the multiplier effect of sustainable development will be set in motion: on the one hand, the amount of resources available will increase, while on the other, the decision-making process concerning their allocation will be more closely in line with the democratic say of the people. Those who are able to satisfy a growing proportion of their local needs through the efficient use of their own resources will also be more able to establish autonomous relationships with the outside world. This concept of *sustainable proximity* will enable the local community to attract the interest of distant investors, and to negotiate with them from a position of independence.

The million-dollar question for the twenty-first century is just what type of globalization will come out on top: that which transfers decision-making to the increasingly distant, autocratic and ungovernable centres of power, or that which is based on local democratic autonomy and which serves to strengthen it. The former seems destined to lead the world down the path of destruction, with the bursting of the giant bubble of speculation. Only the latter can lead us to new forms of ecologically sustainable human development.

References

Acebillo, J. and Folch, R. (eds) (2000) *Atles Ambiental de l'Àrea de Barcelona. Balanç de recursos i problemes*, Barcelona: Barcelona Regional/Ariel, pp. 3–5 and 22–23.

Alió, Mª A. and Olivella, M. (1999) *Per viure bé nosaltres i les generacions que vindran. Com prendre part a fer sostenibles els nostres pobles i ciutats*, Barcelona: Diputació de Barcelona.

Barracó, H., Parés, M., Prat, A. and Terradas, J. (1999) *Barcelona 1985–1999. Ecologia d'una ciutat*, Barcelona: Ajuntament de Barcelona, pp. 66–71.

Botella, J. (coordinator) (1999) *La ciutat democràtica*, Barcelona: Patronat Flor de Maig/Ediciones del Serbal.

Brown, L. R. *et al.* (2000 and 2001) *La situación del mundo. Informe Anual del WorldWatch Institute*, Barcelona: Icaria.

Ciuffini, F. M. (1991) *Proposition de Recherche pour une ville sans voiture*, Rome: Tecnoser.

Ciuffini, F. M., (1993) "El sistema urbà i la mobilitat horitzontal de persones, matèria i energia", *Medi Ambient. Tecnologia i Cultura*, No. 5, pp. 42–53.

Dienel, P. C. and Harms, H. (2000) *Repensar la democracia. Los Núcleos de Intervención Participativa*, Barcelona: Patronat Flor de Maig/Ediciones del Serbal.

Esquerrà, J., Oltra, E., Roca, J. and Tello, E. (1999) *La fiscalitat ambiental a l'àmbit urbà: aigua i residus a la regió metropolitana de Barcelona*, Barcelona: Publicacions de la Universitat de Barcelona.

European Commission (1990) *Libro Verde sobre el Medio Ambiente Urbano*, COM(90)218final, Brussels, p. 30.

European Commission (1997): *Hacia una política urbana para la Unión Europea*, COM(97)197 final, Brussels.

Fernández Durán, R. (1993) *La explosión del desorden. La metrópoli como espacio de la crisis global*, Madrid: Fundamentos.

Fernández Durán, R. (1997) "Movilidad motorizada, globalización económica y 'proyecto europeo'", *Ecología Política*, No. 14, pp. 105–14.

Flavin Ch. *et al.*, (2001) *La situación del mundo. Informe Anual del WorldWatch Institute*, Barcelona: Icaria.

Fòrum Cívic Barcelona Sostenible (1998) *Indicadors de sostenibilitat*, Barcelona: FCBS/Diputació de Barcelona (www.globaldrome.org/FCBS).

García Espuche, A. (1998) "Recuperar la ciutat compacta", *Medi Ambient. Tecnologia i Cultura*, No. 22, pp. 27–35.
Genro, T. and de Souza, U. (2000) *El Presupuesto Participativo: la experiencia de Porto Alegre*, Barcelona: Patronat Flor de Maig/Ediciones del Serbal.
Harvey, D. (1998) "Perspectives urbanes per al segle XXI", en VV.AA., *La ciutat. Visions, anàlisis i reptes*, Girona: Ajuntament de Girona/Universitat de Girona, pp. 113–30.
Hewitt, N. (1998) *Guía Europea para la Planificación de las Agendas 21 Locales. Cómo implicarse en un plan de acción ambiental a largo plazo hacia la sostenibilidad*, Bilbao: ICLEI/Bakeaz.
ITEMA (1996) *Valoración de las emisiones de los gases causantes del incremento del efecto invernadero*, Terrassa: Universitat Politècnica de Catalunya.
Keating, M. (1996) *Agenda 21. Una versió en llenguatge senzill de l'Agenda 21 i els altres acords de la Cimera de la Terra*, Barcelona: Editorial Mediterrània/Fundació Terra, p. 58.
King, A. and Schneider, B. (1991) *La primera revolución mundial*, Plaza y Janés, Barcelona.
Krier, L. (1993) "La civilització industrial davant el repte d'una ciutat nova", *Medi Ambient. Tecnologia i Cultura*, No. 5, pp. 36–41.
Krugman, P. (1997) *El internacionalismo "moderno". La economía internacional y las mentiras de la competitividad*, Barcelona: Crítica.
Martínez Alier, J. (1999) "100 años después de Ebenezer Howard. Economía ecológica y planificación urbana", *Ecología Política*, No. 17, pp. 51–4.
Ministry of Development (2000) *La desigualdad urbana en España*, Madrid: Dirección General de Programación Económica y Presupuestaria.
Naredo, J. M. and Valero, A. (eds) (1999) *Desarrollo económico y deterioro ecológico*, Madrid: Fundación Argentaria/Visor.
Nello, O. (1998a) "La ciutat il·limitada i la ciutat futura", en VV.AA., *La ciutat. Visions, anàlisis i reptes*, Girona: Ajuntament de Girona/Universitat de Girona, pp. 47–73.
Nello, O. (1998b) "Reflexions: el futur de Barcelona", *Medi Ambient. Tecnologia i Cultura*, No. 22, pp. 11–25.
Nello, O., Recio, A., Solsona, M. and Subirats, M. (1998) *La transformació de la societat metropolitana. Una lectura de l'Enquesta sobre condicions de vida i hàbits de la població de la Regió Metropolitana de Barcelona*, Barcelona: Institut d'Estudis Metropolitans de Barcelona/Àrea Metropolitana de Barcelona/Diputació de Barcelona.
Nieto, J. and J. Santamarta, J. (2000) "Evolución de las emisiones de gases de invernadero en España", *World·Watch*, No. 12, pp. 58–61.
Norgaard, R. B. (1997) "Globalización e insostenibilidad", en Universitat Politècnica de Catalunya edit, *¿Sostenible? Tecnología, desarrollo sostenible y desequilibrios*, Barcelona: Icaria, pp. 175–93.
O'Meara, M. (1999) "Una nueva visión para las ciudades", in L. Brown *et al.*, *La situación del mundo 1999*, Barcelona: Icaria, pp. 253–86.
Pindado, F. (2000) *La participación ciudadana en la vida de las ciudades*, Barcelona: Patronat Flor de Maig/Ediciones del Serbal.
PNUD (2000) *Informe sobre el desenvolupament humà*, Barcelona: Associació per les Nacions Unides de Catalunya/càtedra UNESCO de la UPC/Centre UNESCO de Catalunya/Creu Roja de Catalunya.
PNUMA (2000) *Perspectivas del Medio Ambiente Mundial*, Ediciones Mundi-Prensa, Barcelona.
Pruna, I. (coordinator) (1999) *Eines per a una gestió municipal cap a la sostenibilitat. La pràctica diària de l'Agenda 21 Local*, Barcelona: Xarxa de Ciutats i Pobles cap a la Sostenibilitat/Diputació de Barcelona.
Ralea, F. and Prat, A. (1999) *La petjada ecològica de Barcelona. Una aproximació*, Barcelona: Ajuntament de Barcelona.
Rees, W. and Wackernagel, M. (1996) *Our Ecological Footprint. Reducing Human Impact on the Earth*, Gabriola Island: New Society Pub.
Rees, W. (1996) "Indicadores territoriales de sustentabilidad", *Ecología Política*, No. 12, pp. 7–41.
Rees, W. (2000) "Eco-footprint analysis: merits and brickbats", *Ecological Economics*, No. 32, pp. 371–74.

Rueda, S. (1995) *Ecologia urbana. Barcelona i la seva regió metropolitana com a referents*, Barcelona: Beta editorial, pp. 70–71.
Rueda, S. *et al*, (1998) *La ciutat sostenible/La ciudad sostenible/The sustainable city*, Barcelona: Centre de Cultura Contemporània de Barcelona/Diputació de Barcelona, pp. 71–82.
Santamarta, J. (2000) "Los agujeros del cambio climático", *World·Watch*, No. 12, pp. 62–65.
Sen, A. (1999) *Desarrollo y libertad*, Barcelona: Planeta.
Sureda, V. (1999) "Auditoria 21", *Sostenible*, No. 7, pp. 8–12.
Sutcliffe, B. (1995) "Desarrollo frente a ecología", *Ecología Política*, No. 9, pp. 27–49.
Tello, E. (1996) "*Barcelona Estalvia Energia*: una propuesta de democracia participativa para el cambio de modelo de ciudad", *Ecología Política*, No. 11, pp. 43–56.
Tello, E. (1999) "Fiscalitat ambiental i nova cultura de l'aigua", *Medi Ambient. Tecnología i Cultura*, No. 25, pp. 27–39.
Tello, E. (2000) "Los próximos veinticinco años del movimiento ecologista (y los anteriores)", in E. Grau and P. Ibarra (eds), *Una Mirada sobre la red. Anuario de movimientos sociales*, Barcelona: Fundación Betiko/Icaria, pp. 221–46.
Tello, E. (unpublished) *La dinàmica socioecològica del Baix Maresme als anys noranta: l'onada residencial I els seus impactes ambientals i socials*, Innova, Mataró.
Wackernagel, M (1996) "¿Ciudades sostenibles?", *Ecología Política*, No. 12, pp. 43–50.
Wackernagel, M. et al. (2000) *Ecological Footprints of Nations. How Much Nature Do They Use? How Much Nature Do They Have?*, Xalapa, Mexico: Centro de Estudios para la Sustentabilidad/Universidad Anahuac de Xalapa (wackernagel@progress.org).
Wackernagel, M. and Silverstein, J. (2000) "Big things first: focusing on the scale imperative with the ecological footprint", *Ecological Economics*, No. 32, pp. 391–94.
Wallace, P. (2000) *El seísmo demográfico*, Siglo XXI, Madrid, pp. 74–97 and 215–16.
Whitelegg, J. (1993) *Transport for a Sustainable Future. The case for Europe*, London: Belhaven Press, p. 160.
Weizsäcker, E. U. von, Lovins, L. H. and Lovins, A. (1997) *Factor 4. Duplicar el bienestar con la mitad de los recursos naturales. Informe al Club de Roma*, Barcelona: Galaxia Gutemberg/Círculo de Lectores.
World Resources Institute (2000) *Recursos Mundiales. La Guía Global del Medio Ambiente*, Madrid: Ecoespaña.

Notes

1   The original meaning of the word *commuter* was a person who combines two or more means of transport in his daily journey to work, with the corresponding modal interchange (commute), although it has now acquired the accepted definition of someone who has to travel long distances, usually by car, to and from the workplace.
2   With a joint yield from the library's solar system, supplying both photovoltaic electricity and air conditioning, the surface area needed to provide all the domestic electricity consumed in 1996 in Barcelona would be around 1,821 hectares, equivalent to 31% of its constructed surface area. To supply all its needs would require a surface area roughly equivalent to that which today is occupied by all the buildings and their inner courtyards (5,790 hectares). But this is somewhat misleading, because the solar unit installed in the Pompeu Fabra library does not only exploit the roof but also the walls. A considerable area could therefore comprise vertical surfaces. As well as electricity, this system could replace the combustible fuels that are used today in heating buildings.
3   The *Car Free Cities Network* has its head office in Brussels (Place Meeûs 18, B-1050, tel. ++32-2-5520874, fax ++32-2-5520889, email: cfc@eurocities.be). Its website can be consulted via the Agenda 21 for Bremen: www.bremen.de/info/agenda21/carfree/cfcclub.htlm.
4   According to the figures presented by the City Council in 1994 to the Environment and Sustainability Council in the document that forms the basis for the future Local Agenda 21.
5. Cited in the edition of *Nature* dedicated to the Kyoto Summit ("What's new in Nature", http://nature.com).

6. According to data published in the latest issue of the magazine *Sostenible*, published on the Internet (No. 7, Winter 1999/2000, p. 2). See the web page: www.diba.es/xarxasost.
7. See the manual edited on the Internet expressly addressing "the day-to-day practice of the Local Agenda 21" in the town Councils, coordinated by I. Pruna, 1999.
8. See the guide "for participating in the Local Agenda 21", coordinated by Mª A. Alió and M. Olivella, edited by the Xarxa. Also of interest are the books co-published by Patronat Flor de Maig of the Barcelona Council and the Serbal publications in the collection *Res publica* encouraging local experimentation in participative democracy: J. Botella (ed.), 1999; T. Genro and U. de Souza, 2000; F. Pindado, 2000; P. C. Dienel and H. Harms, 2000.

[Translation by Lain Robinson]

Bibliography

The following listing includes only books and articles in English, or with partial or full English translation. No attempt has been made to list Internet resources. The Barcelona City Council web pages, www.bcn.es, have ever-larger parts in English and are in an increasingly useful reference source, including information on many current projects, and extensive statistical data.

Aibar, E. and Bijkar, W. (1997) "Constructing a city: The Cerda Plan for the extension of Barcelona", *Science Technology and Human Values*, 22(1): 3–30.
Ajuntament de Barcelona (1987) *Barcelona: Spaces and Sculptures 1982–86*, Barcelona: Ajuntament de Barcelona.
Ajuntament de Barcelona (1991a) *Barcelona: Urban Spaces 1981–91*, Barcelona: Ajuntament de Barcelona.
Ajuntament de Barcelona (1991b) *Arees de Nova Centralitat/New Downtowns in Barcelona*, 2nd edn, Barcelona: Ajuntament de Barcelona.
Ajuntament de Barcelona (1991c) "Ciutat Vella: The Decisive Moment", *Barcelona: Metropolis Mediterranea* 18, central dossier, Barcelona: Ajuntament de Barcelona.
Ajuntament de Barcelona (1994a) *Barcelona al Mon/Barcelona en el Mundo/Barcelona in the World*, English/Castilian version, Barcelona: Ajuntament de Barcelona, Gabinet de Relacions Exteriors.
Ajuntament de Barcelona (1994b) *Barcelona New Projects*, catalogue for exhibition held in the Salo del Tinell, 29 March–25 May 1994.
Ajuntament de Barcelona (1996a) *Barcelona: Urban Spaces 1981–96*, Barcelona: Ajuntament de Barcelona.
Ajuntament de Barcelona (1996b) *Barcelona: la Segona Renovacio*, Barcelona: Ajuntament de Barcelona.
Ajuntament de Barcelona (1998) *Barcelona: 4 Visions*, Barcelona: Ajuntament de Barcelona (includes text in English).
Amoros, M. (1996) "Decentralisation in Barcelona", *Local Government Studies*, 22(3): 90–110.
Anon (1985) "Homage to Barcelona: the city and its art" (catalogue of an exhibition at the Hayward Gallery, London, 14 Nov. 1985 – 23 Feb. 1986), London: Arts Council of Great Britain.
Anon (1988) *Transformations of a seafront: Barcelona, the Olympic Village 1992*, Barcelona: Gustavo Gili.
Anon (1990) *"Barcelona design guide: Barcelona itineraries for designers, architects and others"*, Barcelona: Gustavo Gili.
Anon (1991) *Barcelona: city and architecture 1980–1992*, Barcelona: Gustavo Gili.
Anon (1994) "Public spaces in the Old Port", *ON Diseno*, 153: 98–103 (in Spanish and English).
Apgar, G. (1991) "Public art and the remaking of Barcelona", *Art in America*, 79(2): 108–21.
Area Metropolitana de Barcelona (1996) *Dinamiques Metropolitanes a l'Area i la Regio de Barcelona*, Area Metropolitana de Barcelona: Mancommunitat de Municipis (English translation included).
Balcells, A. (1996) *Catalan Nationalism: Past and Present*, Basingstoke: Macmillan.
Balfour, S. (1989) *Dictatorship, Workers, and the City: Labour in Greater Barcelona since 1939*, Oxford: Clarendon.

Ball, C. (1990) "Hosting or just taking part?", *The Planner*, 76(34), 31 August 1990: 11–14.

Barcelona Regional (2002) *Barcelona Regional. Agencia Metropolitana de Desenvolupament i d'Infrastructures SA*, Barcelona: Barcelona Regional (full English translation).

Bartolucci, M. and Villalobos, E. (1993) "Barcelona", *Metropolis*, 13(2): 29–35.

Blakeley, G. (2002) "Decentralisation and citizen participation in Barcelona", in P. McLaverty (ed.) *Public Participation and Innovations in Community Governance*, Aldershot: Ashgate.

Blumenfeld, H. (1993) "L'exercise du plan en Europe: Lecture comparee", *C. de L'AURIF*, 104–5: 45–72, August 1993 (11 plans including Barcelona).

Bohigas, O. (1994) "Barcelone, la mer retrouvee", Rennes, Ecole d'Architecture, 1994, pp. 103–6, *Reading and design of the physical environment*, report of a seminar at Rennes School of Architecture April 1993.

Borja, J. (1992) "Eurocities — a system of major urban centers in Europe", *Ekistics*, 352–3: 21–7.

Borja, J. (1996) *Barcelona: An Urban Transformation Model, 1980–1995*, Quito: Urban Management Program.

Borja, J. and Castells, M. (1997) *Local and Global: the Management of Cities in the Information Age*, London: Earthscan.

Botella, J. (1995) "The political Games: Agents and strategies in the 1992 Barcelona Olympic Games", in M. de Moragas and M. Botella (eds) *The Keys to Success: the Social, Sporting, Economic and Communications Impact of Barcelona '92*, Barcelona: Centre d'Estudis Ohmpics i de l'Esport/Universitat Autonoma de Barcelona.

Botti, M. *et al.* (1993) "Barcelona after the Olympics", *Lotus International*, 77: 106–31 (eight articles, in English and Italian).

Bradley, K. (1996) "The great socialist experiment", *Art in America*, February 1996: 72–7.

Breton, F., Clapes, J., Marques, A. and Priestley, G. (1996) "The recereational uses of beaches and consequences for the development of new trends in management: The case of the Metropolitan Region of Barcelona", *Ocean and Coastal Zone Management*, 32, 3: 153–80.

Brunet, F. (1995) "An economic analysis of the Barcelona '92 Olympic Games: resources, financing and impact", in M. de Moragas and M. Botella (eds) *The Keys to Success: the Social, Sporting, Economic and Communications Impact of Barcelona '92*, Barcelona: Centre d'Estudis Olimpics i de l'Esport/Universitat Autonoma de Barcelona.

Buchanan, P. *et al.* (1992) "Urbane village: Nova Icaria", *Architectural Review*, 191: 1146, August 1992.

Busquets, J. (1999) "Spanish waterfronts", *Aquapolis*, 3–4, 50–56.

Busquets, J. *et al.* (2003) *The Old Town of Barcelona. A Past with a Future*, Ajuntament de Barcelona (text in Catalan, Spanish, English).

Catterall, B. (1997) "Citizen movements, information and analysis. An interview with Manuel Castells", *City*, 7: 140–55.

CCCB (1999) *Contemporary Barcelona*, Barcelona: Centre de Cultura Contemporanea (exhibition catalogue with English translation).

Chalkley, B. and Essex, S. (1999) "Urban development through hosting international events: a history of the Olympic games", *Planning Perspectives*, 14, 4: 369–94.

Chalkley, B., Jones, A., Kent, M. and Sims, P. (1992) "Barcelona: urban structure of an Olympic City", *Geography Review*, 6: 2–6.

Dent Coad, E. (1990) *Spanish Design and Architecture*, London: Studio Vista.

Dent Coad, E. (1991) *Javier Mariscal: Designing the New Spain*, London: Fourth Estate.

Dent Coad, E. (1995) "Designer culture in the 1980s: the price of success", in H. Graham and J. Labanyi (eds) *Spanish Cultural Studies: an Introduction*, Oxford: Oxford University Press, pp. 376–80.

Diez, J. (2002) "Metropolitan innovation systems: A comparison between Barcelona, Stockholm and Vienna", *International Regional Science Review*, 25, 1: 63–85.

Dollens, D. (1993) "Post Olympic Barcelona", *Sites*, 25: 72–7.

Dunford, M. and Kafkalas, G. (1992) "The global-local interplay, corporate geographies and spatial development strategies in Europe", in M. Dunford and G. Kafkalas (eds) *Cities and Regions in the NewEurope*, London: Belhaven.

Espuche, A. *et al.* (1991) "Modernisation and urban beautification: the 1888 Barcelona Worlds' Fair", *Planning Perspectives*, 6, 2: 139–59.

Font, A., Llop, C. and Vilanova, J. M. (1999) *La Construccio del Territori Metropolita. Morfogenesi de la Regio Urbana de Barcelona*, Barcelona: Area Metropolitana de Barcelona, Mancommunitat de Municipis (English translation included).

de Forn, M. (1992) "Barcelona: development and industrialisation strategies", *Ekistics*, 352–3: 65–71.

Frampton, K. (1987) "The renewal of Barcelona: an appreciation", in Ajuntament de Barcelona/Joan Miro Foundation, *Barcelona: Spaces and Sculptures (1982–1986)*: 19–24.

Garcia, M. (2003) "The case of Barcelona", in W. Salet, A. Thornley and A. Kreukels (eds) *Metropolitan Governance and Spatial Planning*, London: Spon: 337–58.

Garcia-Ramon, M. D. and Albet, A. (2000) "Pre-Olympic and post-Olympic Barcelona, a 'model' for urban regeneration today?", *Environment and Planning A*, 32, 8: 1331–4.

Giovanni, J. (1992) "Olympic overhaul", *Progressive Architecture*, 73, 7: 62–9.

Graham, H. and Sánchez, A. (1995) "The politics of 1992", in H. Graham and J. Labanyi (eds) *Spanish Cultural Studies: an Introduction*, Oxford: Oxford University Press: 406–18.

Granados, V. (1995) "Another mythology for local development? Selling places with packaging techniques: A view from the Spanish experience with city strategic planning", *European Planning Studies*, 3, 2: 173–87.

Guibernau, M. (1997) "Images of Catalonia", *Nations and Nationalism*, 3, 1: 89–111.

Hargreaves, J. (1999) "Spain divided: The Barcelona Olympics and Catalan nationalism", in J. Sugden and Barner, A. (eds) *Sport in Divided Societies*, CSRCE Vol. 4, Aachen: Meyer and Meyer.

Hargreaves, J. (2000) *Freedom for Catalonia: Catalan Nationalism, Spanish Identity and the Barcelona Olympic Games*, Cambridge: CUP.

Harvey, D. (1989) "From managerialism to entrepreneurialism: the transformation in urban governance in late capitalism", *Geografiska Annaler*, 71 B(1): 3–17.

Hughes, R. (1987) "The spaces and sculptures", in *Barcelona: Spaces and Sculptures (1982–1986)*, Ajuntament de Barcelona/Joan Miro Foundation.

Hughes, R. (1992) *Barcelona*, London: Collins Harvill.

Jauhiainen, J. (1992) "Culture as a tool for urban regeneration: the case of upgrading the Barrio el Raval of Barcelona Spain", *Built Environment*, 18, 2: 90–99.

Jauhiainen, J. (1995) "Waterfront redevelopment and urban policy: the case of Barcelona, Cardiff and Genoa", *European Planning Studies*, 3, 1: 3–23.

Ladron de Guevara, M., Coller, X. and Romani, N. (1995) "The image of Barcelona '92 in the international press", in M. de Moragas and M. Botella (eds) *The Keys to Success: the Social, Sporting, Economic and Communications Impact of Barcelona '92*, Barcelona: Centre d'Estudis Olimpics i de l'Esport/Universitat Autonoma de Barcelona.

Lafaye, J.-J. (1994) "Barcelone: la mer en face", *Connaissance des Arts*, 505: 112–19 (interview with Joan Busquets).

Landry, C. et al. (1996) *The Art of Regeneration: Urban Renewal through Cultural Activity*. Stroud: Comedia.

Leonard, S. (ed.) (1995) "Planning for healthy city regions", *Cities*, 12, 2: 83–115 (section on La Mina housing district — urban stress).

McDonogh, G. (1992) "The geography of emptiness", in R. Rotenberg and G. McDonogh (eds) *Cultural Meanings of Urban Space*, Westport CT: Bergin and Garvey, pp. 3–16.

McDonogh, G. (1997) "Citizenship, locality and resistance", *City and Society Annual Review*: 5–34.

McDonogh, G. (1999) "Discourses of the city: policy and response in post-transitional Barcelona", in S. Low (ed.) *Theorizing the City*, Brunswick, NJ: Rutgers: 342–76.

McNeill, D. (1998) "Writing the New Barcelona", in T. Hall and P. Hubbard (eds) *The Entrepreneurial City: Geographies of Politics, Regime and Representation*, Chichester: John Wiley, pp. 241–52.

McNeill, D. (2001) "Barcelona as imagined community: Pasqual Maragall's spaces of engagement", *Transactions of the Institute of British Geographers*, 26, 3: 340–52.

McNeill, D. (2003) "Mapping the European Left: the Barcelona Model", *Antipode*, 35, 1: 74–94.

Marlow, D. (1992) "Eurocities: from urban networks to a European Urban Policy", *Ekistics*, 352-3: 28–32.

Marshall, T. (1990) "Letter from Barcelona", *Planning Practice and Research*, 5, 3: 25–28.

Marshall, T. (1992) "Of dreams and drains", *Town and Country Planning*, 61, 3: 78–80.

Marshall, T. (1993) "Industry and the Environment in the Barcelona Region", *Town Planning Review*, 63, 4: 349–64.
Marshall, T. (1993) "Environmental Planning for the Barcelona Region", *Land Use Policy*, 10, 3: 227–40.
Marshall, T. (1994a) "Regional Planning and Infrastructure Planning in Catalonia: instruments for environmental gain", in I. Fodor and G. Walker (eds) *Environmental Policy and Practice in Eastern and Western Europe*, Pécs: Centre for Regional Studies, Hungarian Academy of Sciences: 143–59.
Marshall, T. (1994b) "Barcelona and the Delta: metropolitan infrastructure planning and socio-ecological projects", *Journal of Environmental Planning and Management*, 37, 4: 395–414.
Marshall, T. (1995) "Regional Planning in Catalonia", *European Planning Studies*, 3, 1: 25–45.
Marshall, T. (1996) "Barcelona — fast forward? City entrepreneurialism in the 1980s and 1990s", *European Planning Studies*, 4, 2: 147–65.
Marshall, T. (2000) "Urban planning and governance. Is there a Barcelona model?", *International Planning Studies*, 5, 3: 299–313.
Martinez Alier, J. (1996) "The failure of ecological planning in Barcelona", *Capitalism, Nature, Socialism*, 7, 2: 113–23.
Martorell, J. (1992) "The experience of the Vila Olímpica in Barcelona", *Lotus International*, 71: 112–15.
Mendoza, E. (1990) *City of Marvels*, London: Collins Harvill. (Published in Spain as *La Ciudad de los Prodigios*, 1986, Barcelona: Seix Barral.)
Meyer, H. (1999) *City and Port: Transformations of Port Cities London, Barcelona, New York, Rotterdam*, Utrecht: International Books.
Mignot, G. (1991) "Huit centres europeens en course", *C. de l'IAURIF*, 96, April 1991: 53–60.
Miles, M. (2000) "After the public realm: Spaces of representation, transition and plurality", *Journal of Art and Design Education*, 19, 3: 253–61.
Molinari, L. (1992) "Guide to 30 public spaces in Barcelona", *Domus*, 738: 82, May 1992. (Text in Italian, introduction in English.)
Monclus, F.J. (2000) "Barcelona's planning strategies: from 'Paris of the South' to the 'Capital of West Mediterranean'", *Geojournal*, 51, 1–2: 57–63.
Morata, E. (1997) "The Euro-region and the C-6 network: the new politics of sub-national cooperation in the West-Mediterranean area", in M. Keating and J. Loughlin (eds) *The Political Economy of Regionalism*, London: Frank Cass: 292–305.
Moreno, E. and Vazquez Montalban, M. (1991) *Barcelona, Cap a on vas?*, Barcelona: Llibres de L'Index.
Narotzky, V. (2000) '"A different and new refinement'. Design in Barcelona 1960–1990", *Journal of Design History*, 13, 3: 227–43.
Navarro, V. (1997) "The decline of Spanish social democracy 1982–1996", in L. Panitch (ed.) *The Socialist Register 1997*, London: Merlin: 197–222.
Naylon, J. (1981) "Barcelona", in M. Pacione (ed.) *Urban Problems and Planning in the Developed World*, London: Croom Helm: 223–57.
Newman, P. and Thornley, A. (1996) *Urban Planning in Europe: International Competition, National Systems and Planning Projects*, London: Routledge.
Parcerisa, J. (1993) "A city planning laboratory", *Quaderns d'arquitectura i urbanisme*, 198, Jan.–Feb. 1993: 36–41 (Periphery as a project design workshop; in Catalan and English).
Riera, P. and Keogh, G. (1995) "Barcelona", in J. Berry and S. McGreal (eds) *European Cities, Planning Systems and Property Markets*, London: Spon: 219–43.
Rieu, M.-T. (1991) "Les grandes metropoles mondiales, etude documentaire 1991, Barcelona", *Villes en Developpement*, Paris: IAURIF.
Russell, J. and Stein, K. (1992) "Spain's year: Barcelona and Seville", *Architectural Record*, 180, 8: 98–99.
Rowe, P. G. (1997) *Civic Realism*, Cambridge, MA: MIT Press.
Sanchez, J.-E. (1997) "Barcelona: the Olympic city", in C. Jensen-Butler *et al.* (eds) *European Cities in Competition*, Aldershot: Avebury: 179–208.
Simmons, J. and Kamikihara, S. (1998) "Barcelona: commercial structure and change", *Progress in Planning*, 50: 217–32.

Simson, V. and Jennings, A. (1992) *The Lords of the Rings: Power, Money and Drugs in the Modem Olympics*, London: Simon and Schuster.

Smets. M, and Sola Morales, M. de (1994) "The capacity for assessment", *Archis*, 9, September 1994: 50–63 (on laboratory of urbanism approach to design, with example of the Illa; in Dutch and English).

Solokoff, B. (1990) "Public spaces and the reconstruction of the city: learning from Barcelona and Berlin", *Architecture et Comportement*, 6, 4, December 1990: 339–56.

Techniques et Architecture (1996), Dossier on Barcelona, 426: 21–89.

Thompson, J. (1995) "Streets ahead", *Building*, 260, 7881(8), Feb 24 1995: 42–46.

Tolosa, E. and Romani, D. (1996) *Barcelona Sculpture Guide*, Barcelona: Actar.

Torres i Capell, M. (1992) "Barcelona: Planning Problems and Practices in the Jausseley era 1900–30", *Planning Perspectives*, 7, 2, April 1992: 211–33.

Valera, S and Guardia, J. (2002) "Urban social identity, and sustainability — Barcelona's Olympic Village", *Environment and Behavior*, 34, 1: 54–66.

Varley, A. (1992) "Barcelona's Olympic Facelift", *Geographical Magazine*, July 1992: 20–24.

Vazquez Montalban, M. (1992) *Barcelonas*, London: Verso.

Winkelbauer, T. (1992) "Social housing in Poblenou", *Deutsche Bauzeitung*, 126, 3: 46–51 (in German, English summary).

Woodward, C. (1992) *Barcelona,* Manchester: Manchester University Press.

Wynn, M. (1984) "Spain" in M. Wynn (ed.) *Planning and Urban Growth in Southern Europe*, London: Mansell.

Index

Figures and tables in italic, n after page number indicates material in notes

Aalborg Charter 232–3
Agenda 21
  local Agenda 21 232
  movement of cities and towns for sustainability 231–2
  national, come to nothing 232
airport 67, 78, 179–80
architects
  new generation of 54
  and urban change post-1975 8
  well-known, working in Barcelona 211
architecture
  and design of public spaces 94–5
  major component of the city project 138–9
  preservation of architectural heritage 138
  not an object but a process (Lefebvre) 222–3n
  as a project for the city 96
  and the quality of a city 138
  used to disseminate politico-symbolic meaning of cultural space 215
areas of new centrality project 124–7, 144
  areas proposed 125–6
  new developments not aimed at benefiting local people 213
  Olympic areas considered as part of the project 125
  Port Vell *126*, *127*
  project implicit in PGM 125, 126–7
  spatial condition a common factor 126
Associaciones de Vecinos
  election victories 58
  withered away 59
Athens Charter 92

Barcelona 2000 Strategic Plan 75–9, 109
  airport 78
  C-6 city network 77–8
  framework for a new pact with the surroundings 77
  main areas of agreement 76
  population density and use of available space 76–7
  Port–Airport–Mercabarna–Zona Franca–Trade Fair–Logistics Zone complex 78, 175
  public-private sector cooperation 103
  rail network 78
  results of early strategic plan meetings 76
  taking advantage of its high population density 77
Barcelona
  1966 Regional Plan (Pla director) 115–16
  achievements and prospects 20–2
  action plans 114
  beauty may distract from other less satisfactory issues 211
  behind the success story 47–61
  bring value to periphery, recover the centre 113, 114, 212
  challenges to future governability 103–5
  comparisons with the Baix Maresme 242–4, *242*, *243*
  consolidation as capital of Catalan nation without a state 214–15
  costs of being a capital city 79–81
  culture remains Mediterranean 227
  culture, urbanism and the production of ideology 211–20
  factors contributing to integration process 105–6
  fragmentary urbanism, emergence of this style surprising 171
  functioning within a wider context 11–12
  history and geography 2–6
  increased distance between workplace and home 226
  joined "Cities for Climate Protection" 234
  legal framework 74–5
  leisure and tourist site 207–8
  metropolitan policy 108–9
  model challenged 205–24
  municipal government initiatives 97–8
  neighbourhoods came first in overall strategy 60
  new projects summarised 175–90
  now a port-industrial city 212, 219
  people satisfied with local Socialist government 206
  physical change, uneven effects on city region 10
  Pla Macià (1934 plan) 50
  popular consensus to be regarded with scepticism and vigilance 106
  population in the metropolitan region 29–31
  population loss 70, 140–1, 226
  problem of an ageing population 228
  problem of city road traffic emissions 237–8
  public spaces 1980–2000 151–60
  quality of mayor and support of key politicians important 17
  a radical transformation 27–43
  renowned as 'City of Marvels' 49
  resignification and urban transformation 207–11
  schematic facts 9–12
  shortage of land for development 140
  shrinking city 11, *11*, *12*
  social policy included a hardware element 99
  speed of urbanization 225–6
  strategy needed to reduce current ecological footprint 241
  sustainability means a major cut in emissions 239
  urban development, a brief history 49–51
  use of private developers 141
Barcelona City Council
  good administrator of resources 79

257

Barcelona City Council (*continued*)
  greenhouse gas reduction targets *235, 236*
  Metropolitan Programme for the Management of Municipal Waste, well behind schedule 236
  no longer seen as so open to public participation and influence 18
Barcelona Highways Plan, reference point for highway and planning actions 127–8
Barcelona Metropolitan Corporation (CMB: 1974–88) 115, 118
  abolished by Catalan Parliament 128
  acted as planning commission for partial and special urban plans 128
  availability of investment fund from the State 128
  creation of 116
  creation of metropolitan parks 129–30
  creation of road links 130
  creation of wider planning activities 130
  included Barcelona and the municipalities 55
Barcelona model, copied internationally 205–6
Barcelona Saves Energy 234
Besòs river area projects 176–7
  along the Besòs river 177
  along the coastline (waterfront) 177
  Avenida Diagonal 176
  Barcelona—Girona—France rail corridor 176
  new shoreline next to the river 180–4
  Poblenou area 177, 207
Bohigas, Oriol 8
  architect and planning director of the City Council (1980–4) 18
  began by designing and building what was fastest and cheapest 58–9
  important texts 113
  on public spaces actively influencing the urban environment 155
  responded to demands of the neighbourhoods 49
  ten points for an urban methodology 91–6
  urban policy 212
Borja, Jordi
  city, democracy and governability: case of Barcelona 97–110
  public space development in Barcelona 161–2

supported community centre building 17
vital to reform of the council's structure 17
brownfield sites *see* industrial sites, old buildings
  commissioned by the public administration 139
  new housing projects 139
  process of restoration and improvement 135–138

carbon absorption rate, depends on growth of world's forests 239
Catalonia 3, 6
  dependence on nuclear electricity 238
  granted autonomy (1931) 50
  lack of effort by autonomous government 83
  Network of Cities and towns for Sustainability 244
  political stand-off with 100
  population not growing 70
  sustainable culture taking root 244
  wished to show itself a fully fledged nation at the Olympic Games 69
central city, future of 38
Cerdà, Ildefons, 1857 plan for Barcelona (the Eixample) 7–8, *7*, 49, 192
change, four dimensions of 12–30
  politics of change 16–18
  social dimensions of planning 19–20
  treating public spaces and buildings 12–13
  urban changes 13–16
cities
  architectural projects vs. general plans 94–5
  architectural quality 96
  architecture as a project for 96
  as centres of enriching conflict 92
  continuity of the centralities 95
  as domain of the commonality 91–2
  European 91
  have languages of their own 94
  as ideological text 207–11
  Mediterranean and European, fragmentary urbanism an alien concept 171
  metropolitan, organization of 108–9
  movement towards sustainability 232
  old, facing challenges 47
  polarization in the community, ultimate effect of 229
  as political phenomena 91

public space is the city 92
  resignification of 207–8
  "Cities for Climate Protection" 233–4
  joined by Barcelona 234
citizen participation 107
  important in building secure, just cities 105
city, democracy and governability: the case of Barcelona
  challenges to future governability 103–5
  coordination amongst the parties concerned 101–3
  development of public amenities and promotion of the city 98–100
  local identities, multiculturalism and universality 105–6
  the Municipal Charter: a political and legal framework 107–8
  municipal government initiatives 97–8
  organization of the metropolitan city 108–9
  social disintegration and citizen participation 106–7
  urban aesthetics 100–1
Ciudadela Park 168
Ciutat Vella 38, 41
  all not well with current building situation 164
  Area for Integrated Restoration (ARI) 136–7
  better to see continuation than replacement 164
  changes in 85
  joint company formed to improve the district 85
  opening out the urban texture 156
  origins of "opening up" 163–4
  preserving social cohesion and diversity 165–6
  public housing projects 139
  see also Rambla del Raval; Raval neighbourhood
civic patriotism, linked to internationalization of the city 105
civic security and justice, demand for strong 105
Civil War, effects of 4–5
Climate Alliance (Klima Bündnis) 236
  meeting objectives a major step towards sustainability 238
Clos, Joan (Mayor 1997–), lacked interest or expertise 17
CMB *see* Barcelona Metropolitan Corporation

258   INDEX

Coastal Plan 130–1, *132*
  drawing-up began in 1983 130–1
  largest councils carried most weight in decisions 131
collective consumption theory (Castells) 51
Collserola Project 133–4
  area classified as forest park 133, 168
  can define area for unitary management 134
  Collserola Special Plan (PEC) 133
Comarcal Plan 62n
  see also General Metropolitan Plan
commuter syndrome
  could be overcome 245–6
  harmful socio-environmental effects of 229
  solution to 246
compact city, ready to face challenges 38–9
conceived space, space of renovation 152
coordination mechanisms, development of 101–3
cost of living, factor in exodus of people from the city 19–20
Council of Judicial Power, and municipal justice 88
cultural facilities
  Centre de Cultura Contemporània de Barcelona (CCCB) 216–17
  exhibition *The Sustainable City* 229
  key project in transforming the city 217
  Museu d'Art Contemporani de Barcelona (MACBA) 217
cultural homogeneity, and quest for local identity 105
cultural integration 103
cultural spaces, proliferation in Barcelona 211
cultural, technological and entertainment industries
  core of the new economy 213
  see also emerging sectors
culture
  increase in external and internal demand 40
  redefined 214
  see also Forum 2004
culture and leisure centres 41

decentralization
  began with social services 99
  city losing people 30
  consequences of 98
  demographic 30–1
  depends on direct elections 104
  recognizing political pluralism 106
  result of inter-metropolitan residential migrations 31
  strengthened 108
degradation, in the city 113
demographic implosion 228
densification process, added large residential mass 51
design culture, development of from the 1950s onwards 13
Diagonal area 125–6
  inclusion in Olympic Games planning 124
Diagonal Mar transformation projects 143–4, 170, *170*
  large shopping centre proposed 144
  new late-rationalist model 171
  plan is disputed 143–4
  privatization of public space 143, 170–1
Diagonal–Pere IV Park 170
Diagonal–Poblenou transformation project 141–2, *145*
  public-private cooperation 141–2
  see also 22@ area, Poblenou; Poblenou
dispersion
  of people and activities 28
  of urbanization over space (urban sprawl) 30–1
domestic water consumption, increases with income 245

eco-efficiency 234
ecological debt *240*
ecological footprint
  an index of global sustainability 240
  of Barcelona 240, *240*
  and debt, calculation of 239–40
  defined 239
  of the metropolitan expansion 228–9
  shows appropriation of the biosphere by richest people/countries 241
  of various countries *241*
economic growth 33
  growth rate 78
economy
  liberalization at end of Franco's regime 50
  society and environment, changes in 10–11
Eixample 77, 134
  area with regular layout 137–8
  Eixample Ordinance (Bylaw) (1986) 138, 139, 167–8

first plan for Barcelona 49
inner courtyards 167–8
emerging sectors, role of 39–44
  future of the city and Forum 2004 42–4
  new productive activities and 22@ sector 42
  tourist and cultural activity, and new leisure centres 39–41
employers' bodies, supported council's main projects 18
employment, moving out of the city 31
environmental audits 244–5
environmental degradation 232
  continues unabated 233
environmental ministers, recognize failure to stop environmental degradation 232
environmental problem, now not questioned 230
environmental problems
  arise when economy is doing well 234
  arising from urban transformation 34
Estació del Nord Park 169, *169*
European metropolitan system, and Barcelona 70–3
  we are European because we are Catalan 72
Exposicion Universal, showcase for Barcelona 49

fibre optic cabling 77
Forum 2004 14–15, 181, 188, 219–20
  aims 43
  culture as an umbrella word/concept 219
  designed with wider economic objectives 15–16
  a means of promoting the city and restructuring/replanning 220, *220*
  programme for coastal front of Besòs 42–4
Franco, General
  dictatorship not an anomaly 48
  local debate became possible during last years 48
  pursued policy of isolationism and autarchy until 1959 50
  served as visible target for the opposition 51
Franquismo 47
  growth of urban social movements during final years 51
  towards the end urban movements acquired more force 48
  weakening grip of 53
Front Maritim, new housing 143

Gaudí, Antoni 210
General Metropolitan Plan 53, 54–7, *54*, 112–14
   1976 Plan, valid instrument for public change 114–18
   1992, a key reference point in planning 140
   acquisition of much designated public area land by the city 58
   attacked by neighbourhood associations 55–6
   beginning of planning in the public interest 48
   capacity for modification to accommodate local plans 118
   conflicts over caused substitution of new conciliatory mayor 56–7
   derived from planning law 114
   effects on conservation and improvement of city buildings 137
   effects of new generation of architects 54–5
   first version (1974) 48
   implementation begins 48
   land for private initiatives 118
   large land reservations gave space for highway proposals 127–8
   less acceptable concessions amended by central government 117
   major planning projects: search for general reference points 122–8
   metropolitan planning projects 112, 113, 128–34
   restoration and improvement of buildings 134–9
   scope 115
   seen as an obstacle to be overcome (PERIs) 120
   technocratic proposal 117
   urban transformation projects after 1992 112–13, 140–4, *145*
   work of "infiltrated experts" 116
   worst excesses removed by the courts 57
   see also PERIs
gentrification 209, 217
globalization
   economic 230
   growth of transactional costs 246
   new culture of global sustainability 230
Green Book on the Urban Environment 225
greenhouse gases
   direct emissions in Barcelona, increase in *235*

Heidelberg's voluntary commitment 233
   main groups of 234, *235*
   per person per year in Spain 238–9
   reductions in promised at Kyoto 233
Heidelberg
   initiative, voluntary commitment on greenhouse gases 233, 236
   meeting objectives a major step towards sustainability 238
   summit of local authorities, "how to combat global climate change" 233–4
housing
   and the baby boomers 227
   net demand, the young and the elderly 227
   poor go to areas with affordable prices but few amenities 226
   residents per house falling 226
   rising prices, causes of 227–8
   supply for young couples inadequate 104
housing plan
   accessible housing 86
   public housing 87
   reasons for 86
   rehabilitation of sea front with housing 87
hydrological deficit, may require transfer of water 245

Ibero–American Centre of Urban Strategic Development, and the Barcelona Strategic Plan 76
immigration 11
   of impoverished ethnic groups 105
industrial activity, the image of Barcelona 27
industrial land, lucrative revaluation of 213
   led to dislocation of original neighbourhoods 213
industrial sectors, renovation of 144
industrial sites, old
   conversion to parks 153, 168–9
   key source of development land 15
   Poblenou, conversion to residential, productive and service functions 193–4
   uses of 208–9
information, tensions and chance as instruments of 92
inner courtyards, *Eixample* 167–8
integral rehabilitation, an objective for some areas 37

Joan Miró Park 169, *169*
justice system
   decentralized/de-concentrated system 88–9
   peacemaking justice 88

Kyoto protocol 233

La Catalana
   gradually lost population 183
   planned urban renovation work 183–4
land and housing market, rigidities in 32–3
landowners, attacked 1974 General Plan 56
Lefebvre, Henri, three moments of social space 151–2
leisure and shopping centres 162–3
lived space, space of people 152
Llobregat delta
   airport 179–80
   Delta Plan schemes 178
   infrastructure important in establishing business districts 178–80
   port 178–9
Llobregat river area projects 175–6
   environmental improvements 176
Llull-Taulat: new residential/business neighbourhood, to incorporate sustainable development and energy saving 182–3

Manresa Declaration, main objectives 244
Maragall, Pasqual (Mayor 1982–97) 8, 16–17, *16*, 117, 158
   ability to make deals with Madrid 17
   anti-nationalist stand 215–16
   and citizens' perceptions of public space as a common good 210
   and the goal of social justice 155
   governing Barcelona 65–89
   and the Olympic Games 60
   on the proposed transformation of Raval neighbourhood 156
   Refent Barcelona (*Making Barcelona*) 214
   used Olympic Games for major transformations 217, *218*, 219
Maremagnum: private public space 162–3, *163*
   leisure and shopping centre 162–3
   relationship with the Rambla 163
market, taking over from policy 95

Masó, Enric (mayor), attempted to establish dialogue with neighbourhood associations 55
Mediterranean cities 1
  rediscovery of the traditional urban system 225
metropolitan area, wider 20
  equally massive changes 10
  gaining population 226
  greater supply of houses at affordable prices 227
metropolitan planning projects 128–34
  Coastal Plan 130–1, *132*
  Collserola Project 133–4
metropolitan region 29–33
  in an accelerated process of change 37
  functional and social specialization 32–3
  Maragall's position on 71
  not growing 70
  population sprawl and economic activities 30–1
  total population unequally spread 29
  urban area expansion 32
metropolitan structure, pitfalls to be avoided 109
metropolitanization
  benefits of 33–4
  cities integrated into a network 38
  diffuse and socially segregating 228–9
  increases waste generation and greenhouse gas emissions 242, *242*
mobility, unsustainable rise in 230
mobility needs, of citizens and businesses, explosion of 34–6
mobility networks, metropolitan, problems with 36
Montjuïc 125–6
  had to be included in Olympic Games planning 124
  National Palace 67–8, *67*
monumentalization
  objects from past turned into empty shells 209
  of the peripheries 166, 212
  street sculptures 153–4, 223n
Municipal Charter (Special Law) 73–5, 107–8, 110n
  1994 draft less ambitious 108
  Autonoma University, founding of 74
  central municipality, relationship with surrounding municipalities 73
  collaboration of different institutions 73–4
  cultural bodies 73–4

history of 73
  issues of political organization 107–8
  provides for civic (joint) management of major projects 103
municipal government
  campaigns calling for public cooperation 102
  initiatives 97–8

nationalism in post-modernity, different responses to 215–16
neighbourhood associations 106, 154
  demanded changes to plan 57
  opposed 1974 Plan 55–6
  response to problems specific to the neighbourhood 52
  scope expanded in early 1970s 53
neighbourhood protests, intangible network between 52

OECD, and the Barcelona Strategic Plan 76
Olympic Games 17, 38, 60, 66–70, 140, 213, 217–19
  1986–92 investment programme as a model 188–90
  Barcelona's debts 80–1
  cooperation among public administrative bodies 103
  elements of plurality in opening ceremony 69
  image projected 69
  infrastructure, 67–8
  planning projects for 123–4
  seen as a project by all for all 217
  a spur to Barcelona's energy 70
  stimulated tourist activity 40
Olympic holding company 70

parks, new, reclaimed spaces 168–70
  effective method of creating quality spaces 170
  multi-purpose spaces 169
Partial Plans (*Planes Parciales*), opposition to 53
perceived space, spaces of design 151–2
peripheral developments 50–1
  monumentalized 166, 212
  reasons for appearance 95
PERIs (localized planning projects) 112, 114, 119–21, *119*, *122*, 134
  also instruments for real planning 121
  concern to improve the existing city 119–20
  importance beyond that of a simple planning instrument 120

main PERIs *122*
  proposals seen as development of the PGM 121
  sometimes defensive role against the PGM 120
PGM *see* General Metropolitan Plan
planning
  rests on political foundations 18–19
  social dimensions 19–20
  see also town planning
planning policy, aims of (Abad) 154–5
planning projects, major: search for general reference points 122–8
  areas of new centrality 124–7
  Olympic project 123–4
  road network 127–8
planning tradition 6–9
  emergence of strategic planning 9
  insistence on giving physical and visual form to schemes 8–9
  involvement of politicians 8
  planning carried out by engineers and architects 8
  plans for Barcelona 6–9, *7*
plans/programmes 6–9
  architectural projects vs. general plans 94–5
  impacts of 15–16
  metropolitan planning projects 128–34
  Pla Macià (1934 plan for Barcelona) 50
  see also General Metropolitan Plan; 22@ area, Poblenou
Poblenou 67, 177, 193–4, 207
  Diagonal–Poblenou transformation project 141–2, *145*
  see also 22@ area, Poblenou
political and social problems 104
political transformation, Spain, Catalonia and Barcelona 78
population
  global, majority live in urban areas 232
  metropolitan Barcelona 29–31
  stagnating, consuming more land for new homes 226
population density
  in Barcelona 2000 Strategic Plan 76–7
  control of in city by PGM 137
Porcioles, Josep Maria de (Mayor 1957–73), mandate ended by destructive partial plan 53
Port Vell 41
  an area of new centrality, successful but criticised 127
  suggested schemes *126*

INDEX **261**

President of the Autonomous Community, important role for 69
Princep de Girona gardens 170, *170*
private finance
  and public urban development 212–13
  successful in profitable sectors 14
  urban change, sometimes generates tension and debate 13–14
private spaces, pretending to be public spaces 162–3
property purchase, a safe investment 228
proportional representation system, depersonalizing political representation 104
proximity
  participative democracy 231–2
  strategy of (sustainable proximity) 246–7
public amenity development and promotion of the city 98–100
  creation of an attractive environment 98
  promotional campaigns, linked to actual events 99
public cooperation campaigns, municipal government 102
public debt 140
public safety 87–9
  negative response to idea of municipal justice 88
public spaces 38, 211–12, 223n
  Barcelona 1980–2000 151–60
  buildings 12–13
  designed in architectural terms 94–5
  development in Barcelona, examples 161–72
  identity of 93, *93*
  improvements to 9–10
  much investment in peripheral districts 156
  quality and quantity impoverished by new developments 213
  response to urban space 156–7
public transport, investment in 185, *186*
public works 114
public-administration relations 107
public-private cooperation, in business and job creation 102–3
public-private parnerships 14

quality of life in a large city 81–3
  many things not working properly 82
  need for collaboration 82
  a social objective 81

railway, moved from Poblenou 67
Rambla del Raval 156, *156*, 207
an opportunity? 163–6
ramblas, new 166–7
  mixed spaces 166
  Ronda del Mig ring road, covering of 167
  Via Júlia 166
Ramoneda, Josep, connection between modernity and the city 216
Raval neighbourhood
  proposed transformation to a cultural centre 156
  public space used to express social tensions 157–8, *158*
  Rambla del Raval 156, *156*
Raventós, Francesc, discussion on balances and approaches in public operations 14–15
road network 127–8
  interdependence with urban fabric 127
  investment priority over public transport 25–6
Royal Institute of British Architects (RIBA), awarded Gold Medal to Barcelona 60–1, 91, 205

Sagrera — Sant Andreu *see* Sant Andreu–La Sagrera project
Sant Andreu–La Sagrera project 142–3
  internodal transport centre 142
  remodelling of the urban environment and the future central railway station 184–5
  riverside park 142
  siting of TGV station at La Sagrera 142, *142*, 184
Serra, Narcís, Mayor 16, *16*, 58
  began series of important projects 59
  responded to needs of neighbourhoods 58, 59
Serratosa, Albert
  dismissal 116
  engineer for 1974 General Plan 55
  pressured by landowners 57
  on the role of the neighbourhood associations 57
shopping centres, proposed by private developers 144
social cohesion 42, 114
  erosion of 229–30
  maintained, Ciutat Vella 165
social disintegration, and citizen participation 106–7
social disruption
  arising from exclusion 104
  through transformation of old industrial areas 213–14

social housing
  only possible where authorities could subsidize 14
  provision of public facilities neglected 51
social liberties, lack of made organization difficult 52
social polarization 233
social segregation, and the housing market 36–7
social solidarity, an issue, challenge of immigration 21–2
Socias, Josep Maria (Mayor)
  had to be responsive to neighbourhood associations 57
  hired Solans as planning director 58
Soja, Edward, Thirdspace 159
Solans, Joan Antoni, planning director 48, 55, 58, 117
solar radiation
  greater in Maresme 245
  use of a possibility in Barcelona 236–7
spaces of design 152–4
  conversion of industrial sites to parks 153
  importance of new public spaces 152
  important projects completed 153
  reformed spaces 152
spaces (Lefebvre) 151–2
spaces of people 156–8
spaces of renovation 154–6
Spain
  Barcelona and Catalonia 72
  incipient metropolitan system 71
specialization 29
street sculptures 223n
  effectiveness of 153–4
Subirós, Pep, on culture as a key industry 214
suburban developments, residential peripheries 50
sustainability 114, 182–3
  local action gives rise to hope 231
  local, principles and tools for 225–50
  obstacles attached to achieving are enormous 231
  see also unsustainability

telecommunications, investment in 185, *186*
tertiarization 10, 217
tourism 208
  expanding in Barcelona 39–40
town planning
  22@ area, Poblenou 194–5, *195*
  coordinated 102

transport management 83–5
 government interventions, effects of 84–5
 mobility problems in the city centre 83–4
 see also mobility needs; road network
22@ area, Poblenou 14–15, 185
 controversy over plans 19
 modification of General Metropolitan Plan 42, 193–4, 195–6
 new activities to generate new jobs 42
 origins and past 191–2
 influence of Eixample plan 192
 problem, low amount of publicly owned land 15
 processes of controlling change 194–5
 varied landscape 192–3
22@ plan
 @ activities 196–7, 200–1
 conditions for transformation 199–200
 establishes criteria for transformation to new activities 193–4
 extent of the transformation 200
 new buildings alongside reused old buildings 195
 new infrastructure 197–8, *198*
 proposed development planning instruments 198–9
 a regulatory document 194
 special plans 194–5, *195*
 transformation management 197
 uses permitting in new zoning regulations 195–7

unemployment 78
 needs municipal response 104
Universal Forum of Cultures, Barcelona 2004 *see* Forum 2004
unsustainability
 drift towards 228–30
 tracking the eco-footprints of 239–41
urban aesthetics 100–1
 functions of 101
urban changes 13–16
 financing the programmes 13–15
 impacts of the programmes 15–16
urban character, giving continuity to 95
urban development and planning
 criticism 19
 gains for many inhabitants 19
urban environment
 city building activities 100
 enhancement contributed indirectly to economic revival 99, *100*
urban fabric, and the road network 127–8
urban methodology, ten points for 91–6

urban network, rich and compressed, Barcelona 71
urban policies
 defending model of compact, complex and integrated city 37
 question of governance 37–9
 The Urban Question, Manuel Castells 51
urban renewal and development projects, opportunities for cooperation 102
urban renovation 134–5
 combining transformation with memory 38
 reconfiguration and new signifiers 208–9
 urban policies 37–9
urban shopping centres, new 41, 144, 162–3
urban social movements 51–2
 Barcelona case 52–3, *54*
 lost momentum, power and membership 59
urban space
 central role of politics in construction of 39
 identity 93
 legibility 93–4
 public space a response to 156–7
 road network as an integrating factor 127
 text 222n
urban speculation 210
urban sprawl 30–1
 causing environmental problems 34
 extension of 28
 many people now living in low accessibility areas 36, *36*
urban text 207–11
 city and tourism 208
 new monumentality 208–9
 political/ideological dimension to aesthetics 210
urban texture
 Ciutat Vella 156
 public spaces and plots of land between 134
urban transformation
 Barcelona 210–11
 buildings, not reached standard of public spaces 135
 challenges 34–7
 dynamics 28–9
 projects after 1992: recycling urban space 140–4, *145*

urbanism
 culture, linked by creation of public space 211
 fragmentary, alien concept for Mediterranean and European cities 171
 partial and fragmentary 171
 and the production of consensus as interconnected processes 207
 replacement by architecture 94
 socio-ecological 225
urbanization, marginal areas, illegal 50
USA, on "global communism"! 241

Via Júlia, rambla in a new location 166
Viola, Joaquim (Mayor) 116
 friend of Porcioles 57

water, double system proposed (drinking and recycled water) 71
waterfront 41, 177
welfare state, introduced in Spain 37
young people, marginalization of growing 104–5